T0271014

Modelling, Stability Analysis, and Control of a Buck Converter

A generalized approach in a systematic way is inevitable to oversee the challenges one may face in the product development stage to acquire the desired output performance under various operating conditions. This book, *Modelling, Stability Analysis, and Control of a Buck Converter: Digital Simulation of Buck Regulator Systems in MATLAB®*, written and structured to cater to readers of different levels, aims to provide a clear understanding of different aspects of modelling and practical implementation. The operation of the semiconductor switches, switching characteristics of the energy storage elements, stability analysis, state-space approach, transfer function modelling, mathematical modelling, and closed loop control of the buck converter, which are illustrated in this book can be extended to any other similar system independent of complexity.

This book:

- Covers modelling and control of buck converters and provides sufficient understanding to model and control complex systems.
- Discusses step response, pole-zero maps, Bode and root locus plots for stability analysis, and design of the controller.
- Explains time response, frequency response, and stability analysis of the resistive-capacitive (R-C), resistive-inductive (R-L), and R-L-C circuits to support the design of the buck converter.
- Includes simulation and experimental results to demonstrate the effectiveness of closed loop buck regulator systems using proportional (P), integral (I), and P-I controllers to achieve the desired output performance.
- Provides MATLAB codes, Algorithms, and MATLAB/PSB models to help readers with digital simulation.

It is primarily written for senior undergraduate and graduate students, academic researchers, and specialists in the field of electrical and electronics engineering.

Modelling, Stability Analysis, and Control of a Buck Converter

Digital Simulation of Buck Regulator Systems in MATLAB®

Moleykutty George and Jagadeesh Pasupuleti

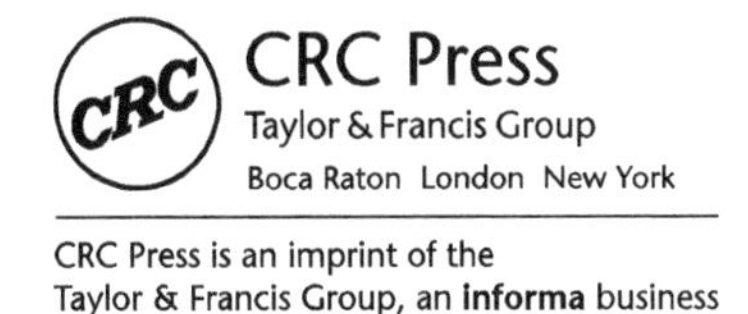
CRC Press
Taylor & Francis Group
Boca Raton London New York

CRC Press is an imprint of the
Taylor & Francis Group, an **informa** business

Designed cover image: Moleykutty George

First edition published 2025
by CRC Press
2385 NW Executive Center Drive, Suite 320, Boca Raton FL 33431

and by CRC Press
4 Park Square, Milton Park, Abingdon, Oxon, OX14 4RN

CRC Press is an imprint of Taylor & Francis Group, LLC

ISBN: 978-1-032-62773-1 (hbk)
ISBN: 978-1-032-84113-7 (pbk)
ISBN: 978-1-003-51123-6 (ebk)

DOI: 10.1201/9781003511236

Typeset in Sabon
by Apex CoVantage, LLC

Dedicated to

My parents, late Mr. and Mrs. Puliyelil Kochumalayil George George for their trust in me to leave their unaccomplished lifetime goals with me, and to my eldest sister Girija Mohan for holding me without falling since childhood with love, affection, and care.

Moleykutty George

Dedicated to

My parents, wife, and daughter.

Jagadeesh Pasupuleti

Contents

Preface

Buck regulator systems (BRS) are used to step down high-level DC voltages to low-level DC voltages and the system should provide regulated output voltages irrespective of the variations in the input, parameters, and the load restricted within the design factors. The Buck regulator system, a standalone device using controllable semiconductor switches operating at high frequency, is introduced to reduce harmonics mitigation and power dissipation and also to minimize the weight and cost.

The aim of this book is to provide a concise illustration of the operation, mathematical modelling, digital simulation, stability analysis, and control of a buck converter circuit. The systematic approach used in this book should enable researchers to model any similar complex system. The book doesn't exclude the practical aspects to be considered during product development stage. The book is written and structured in such a way that readers who are weak in mathematics, circuit theory, and control systems also will still be able to have a clear understanding of different aspects of modelling, control, and practical implementation.

Drawbacks of AC-DC converters and their impacts on power quality are explained in the introduction of this book with the help of MATLAB®/Power System Blockset model (PSB) configurations of single-phase and three-phase AC-DC converters, output voltage waveforms, source current waveform, graphs of total harmonic distortion, and reactive power associated with AC-DC converters. The effectiveness of MATLAB codes/toolboxes for modelling, analysis, and control is demonstrated throughout the book. MATLAB provides a user-friendly environment to model systems that include electrical, electronics, and control systems. Step-by-step procedures to develop Simulink models, conduct simulation, and to retrieve simulation results are also included to help beginners to start with modelling and simulation. The working of semiconductor switches that modernize the world with minimal cost is well illustrated with the help of PSB models and system responses. A PSB model of a buck converter is included in the introduction of this book to explain the working, advantages, and applications of buck converters.

In BRS, unlike resistors, inductors and capacitors store/release energy rather than dissipating energy. Chapter 2 explains the instantaneous power associated with capacitors and inductors in energy storage/release modes with the help of separate circuit configurations. Switching characteristics of capacitors and inductors play an important role in BRS. Illustrations are provided in the second chapter of this book to explain the switching behaviour of a capacitor in a resistive-capacitive (R-C) circuit with variations in frequency. Derivations and tables provided in the second chapter relate the voltage across the capacitor with parameter variations. Plots using conventional approach and PSB simulation are also included in the second chapter to demonstrate the variations in the charging time with RC time constant. State-space approach is used to model the dynamic behaviour of a system. The chapter also details the mathematical formulation of the state-space model of a resistive-inductive (R-L) circuit. MATLAB codes are provided in the second chapter to plot the current through the inductor and the voltage across the inductor of an R-L circuit using state-space approach. Illustrations are also included in the second chapter to demonstrate the output response of an R-L circuit using state-space approach.

Representation of the system gain with respect to exponentially varying sinusoidal waves is used to predict and graph the system behaviour in the Laplace domain. Chapter 3 shows the step-by-step procedure to represent an R-L-C circuit in the Laplace domain; mathematical formulation of the transfer function model of an R-L-C circuit with a specific example is also detailed in the third chapter. The chapter also includes derivations to obtain the poles and zeros from the transfer function model of a system; physical concepts of poles and zeros are explained with the help of circuit diagrams. Pole-zero maps of the R-C circuit for different parameter values included in the third chapter illustrate the relationship between the pole locations and the charging times. Examples are included to verify the pole-zero maps generated using MATLAB codes. One may use a pole-zero map to predict the stability of a system; step responses of the systems with poles at the origin, with complex conjugate poles on the imaginary axis, with complex conjugate poles on the left half of the s-plane, with complex conjugate poles on the right half of the s-plane and with poles/zeros on the right half of the s-plane illustrated in the third chapter show how the location of the poles-zeros affects the stability of a system. Derivations are included in the final part of the chapter to obtain the time response of a system from the transfer function model; MATLAB codes are included to verify the results obtained using partial fraction techniques and inverse Laplace transform.

Chapter 4 includes the advantages of closed loop systems in achieving the desired output response over the open loop systems. Illustrations and examples are provided in the fourth chapter to explain the instability happens in a closed loop system even when the open loop system is stable. The chapter details the variations in the magnitude and phase of the system gain when the system is exposed to sinusoidal waves of constant magnitude but over a wide range of frequency using logarithmic scales namely Bode plot. Calculations and illustrations are provided to discuss the Bode plot of an R-L circuit with parameter variations. As the frequency of the input signal is varied, for certain frequency the system may introduce instability with an inverted output of the same magnitude. Gain margin (Gm) and phase margin (Pm) obtained from the Bode plot indicate to what extend the gain and phase of the system can be adjusted without reaching instability. The chapter explains and verifies the gain margin and phase margin obtained using MATLAB codes. Explanations are given with the help of Gm, Pm, pole-zero maps, and step responses on how the transfer function of an open loop system is used to predict the stability of the closed loop system and how the gain can be varied to achieve stability. The fourth chapter also details the systematic approach of plotting root locus and obtaining different values of gain from the root locus using MATLAB command to analyze the stability of the system for different values of gain.

The operation of a buck converter circuit is discussed in Chapter 5. Operation of the power circuits in two modes, one with the controllable switch ON and the other one with controllable switch OFF, is explained in the fifth chapter with the help of circuit diagrams, examples, and illustrations. Calculations, tables, and illustrations are provided to discuss the performance of the open loop buck converter circuit with variations in the peak-peak ripple current of the inductor, peak-peak ripple voltage of the capacitor, and the switching frequency. Critical values of the inductor and the capacitor are also considered for discussion.

Mathematical models describing a system with sets of equations are used for design, analysis, and control. The state-space model uses first order differential equations to model any dynamic systems. Chapter 6 covers a systematic approach of obtaining the state-space average model of the buck converter, considering the two different modes of operation of the power circuit of the buck converter. The stability of a system would be affected with changes in the input-output conditions. Derivations of large signal, steady state, and small signal average models of buck converters are included in the chapter to analyze the system behaviour under transient, steady state, and minor variations in the operating conditions. Large signal average models and small signal average models of the buck converter included could be used for the stability analysis. For the closed loop system to be stable, the open loop gain should not have any poles/zeros on the right half of the s-plane, and both Gm and Pm should be positive. Pole-zero map, Bode plot, and closed loop step response of the large signal average model and small signal average model of the buck converter are also included in the chapter to analyze the stability of a buck converter circuit.

Performances of the mathematical models derived in Chapter 6 are exclusively verified in Chapter 7; both open loop and closed loop systems are included for the analysis. Transfer function models of the open loop large signal average model and small signal average model have been derived to analyze the time response using partial fraction techniques; initial and final value theorems have also been applied for the analysis. MATLAB codes and SIMULINK models are also included in the chapter to verify the accuracy of the mathematical models. State-space approach is used to analyze the performance of steady state average model. An algorithm is included in the chapter to obtain the steady state output of the model; results achieved are also discussed. The chapter also discusses closed loop control of buck regulator systems; algorithms are provided to determine the gain using the root locus technique. One may use a proportional controller to vary the rise time, an integral controller to vary the steady state error, a derivative controller to vary the overshoot/oscillations, or a combination of the controllers based on the specific output requirements. The control of a buck regulator system using an integral controller is included in the chapter with a large variation in the input voltage. An algorithm, the results obtained thereby, and illustrations included in Chapter 7 demonstrate the successful application of a MATLAB program developed for the control of a buck converter with a large variation in the input voltage. The control of a buck converter using a proportional controller with a small deviation in the duty cycle is also covered. Calculations are included to determine the limiting value of gain factor. Illustrations and results included in the chapter supports the validity of the control technique used for the small signal model.

Chapter 8 covers modelling and simulation of a buck regulator system using PSB. PI-PWM control is used to regulate the output voltage of a buck converter system to the desired value independent of the variations in the load. One may note that the PI controller generates a control voltage given by $v_{con} = k_p error + k_i \int error$ which is then compared in a PWM circuit with a carrier signal switching at high frequency to generate the switching pulses required to drive the controllable switch. Zeigler-Nichols' trial and error method has been used to determine the vales of the proportional constant k_p and the integral constant k_i. PSB model of a BRS, plots of input voltage, desired output voltage, actual output voltage, load current, control voltage, and switching pulses generated are included in Chapter 8 to demonstrate the successful control of the buck regulator system independent of the variations in the load. A case study has also been included to demonstrate the effectiveness of a flip-flop controlled buck regulator system. The chapter discusses the practical aspects to be considered during product development stage. A discussion on the utilization of solar panels to provide input supply for the BRS is also covered. Practical implementation of BRS included in the chapter demonstrates the elimination of costly and heavy weight apparatus as well as complex techniques. The control circuit used for the practical implementation of BRS uses the ability of flip-flops to reproduce the input even at high switching frequency; and the switching signals are originated after comparing the control signals with a realistic fixed reference value. The case study incorporates both high side and low side switching configuration of an n-channel MOSFET. Effective application of the flip-flop-based control circuit for input voltage control and duty cycle control has been included as a part of the case study.

References/bibliography are included in every chapter of this book to address the original work done and for any further reading on a specific topic. Online resources are also provided for readers to nourish understanding on modelling and control. Readers who are new to MATLAB modelling and simulation are encouraged to refer and practice examples/demos provided with the referred software package. Textbooks on circuit theory, switch mode power supplies, and control systems are also referred to in this book for readers to explore various aspects of a specific subject matter.

About the authors

Moleykutty George is the founder of Rev. George Mathan Research Centre, Malaysia. Her areas of research are in the fields of switch mode power conversion, power quality improvement, and speed control of electric drives. Publications of Dr. George mostly emphasize the applications of artificial intelligent controllers in the field of electrical and electronics engineering. She has been in the academic field since 1992 with a total of 16 years of teaching experience at the University level and 12 years of experience as a full-time researcher.

Jagadeesh Pasupuleti is the Professor and Head of Hybrid Renewable Energy System, Universiti Tenaga Nasional, Malaysia. Prof. Ts. Dr. Jagadeesh Pasupuleti has 35 years of experience, supervised 35 postgraduate students, involved in 60 funded research/consultancy projects and published 150 papers in international conferences/journals. Prof. Ts. Dr. Jagadeesh Pasupuleti is a reviewer, session chair, international program committee member, key-note speaker, senior Member of IEEE, Member of IET, chartered engineer, professional review interviewer for CEng, member of EI (UK), member of BEM (Malaysia), member of MBOT (Malaysia), professional technologist (Malaysia), and member of ISTE (India).

Acknowledgements

First, the authors would like to praise and glorify almighty GOD for blessing us with wisdom, knowledge, guidance, strength, faith, confidence, and peace for the successful publication of this book.

The authors are grateful to the anonymous reviewers who made valuable suggestions at the early stage of the revision process.

The authors extend their sincere gratitude to our publisher for timely correspondence, constructive suggestions, and co-operation.

The authors acknowledge with thanks the direct/indirect help, support, and contribution of the following organizations/academics/friends for the successful publication of this book:

- Multimedia University
- Universiti Tenaga Nasional
- NPTEL, Ministry of education, Government of India
- Prof. L. Umanand
- Prof. K. N. Pavithran
- Prof. Saurabh Kumar Mukerji
- Prof. Kartik Prasad Basu
- Prof. Nirod Chandra Sahoo
- Prof. Bala Venkatesh
- Prof. Suresh Kumar K.S.
- Mrs. & Mr. Christo George
- Prof. Ned Mohan
- Prof. Muhammad Rashid
- Prof. Adbulazeez S. Boujarwah
- Prof. Vineetha Kalavally
- Prof. Mini Sreejeth
- Mdm. Juno Nesamony
- Datin Julie K. Cherian

The authors express their heartfelt thanks and love to their family members for their continuous support, encouragement, understanding, and patience shown throughout this book project.

The authors are also grateful to their beloved students for timely feedback throughout their academic career without which this book may not have been a reality.

Last but not least, with gratitude, the authors value the timely assessment of the critics for continuous improvement throughout their academic journey.

Chapter 1

Modelling of a buck converter

An overview

1.1 INTRODUCTION

With the advent of fast acting fully controllable semiconductor switches, switch mode power supplies (SMPS) were introduced to overcome the demerits of typical AC-DC power supply. SMPS transforms DC voltage from one level to another and regulates the DC output voltage. Commonly used SMPS topologies are buck, boost, and buck-boost, based on the way the switches and energy storage elements are configured within the device. A buck converter configuration is used to step down the DC voltage, a boost converter to step up the DC voltage, and a buck-boost to step up or step down the DC voltage.

1.2 DRAWBACKS OF AC-DC CONVERTERS

1.2.1 Single-phase AC-DC converter

Consider a typical AC-DC converter with system specifications as seen in Table 1.1; as shown in Figure 1.1, the typical AC-DC converter requires a transformer in addition to the semiconductor switches; and eventually this increases the power consumption, power dissipation, weight, size, and cost. One may also note that the output of the AC-DC converter is pulsating in nature as shown in Figure 1.2 and thus requires a filter circuit and a voltage regulator to produce a regulated DC output voltage.

1.2.2 Three-phase AC-DC converter

Figure 1.3 shows the circuit configuration of a three-phase AC-DC converter with the system specifications as seen in Table 1.2. As shown in Figure 1.4, the source current wave form of the three-phase AC-DC converter is not purely sinusoidal in nature and is harmonic polluted; one may note that as shown in Figure 1.5, the total harmonic distortion is about 44% which is above the IEEE-519 standard of 5%. Analysis of Figure 1.6 shows that the AC-DC converter also introduces reactive power into the power supply mains.

DOI: 10.1201/9781003511236-1

Table 1.1 System Specifications of the Single-Phase AC-DC Converter

Input		*Transformer T_1*		*Load resistance*
V_s (V)	f_s(Hz)	VA	N_1/N_2	R_{dc} (Ω)
120	60	100	120/24	5

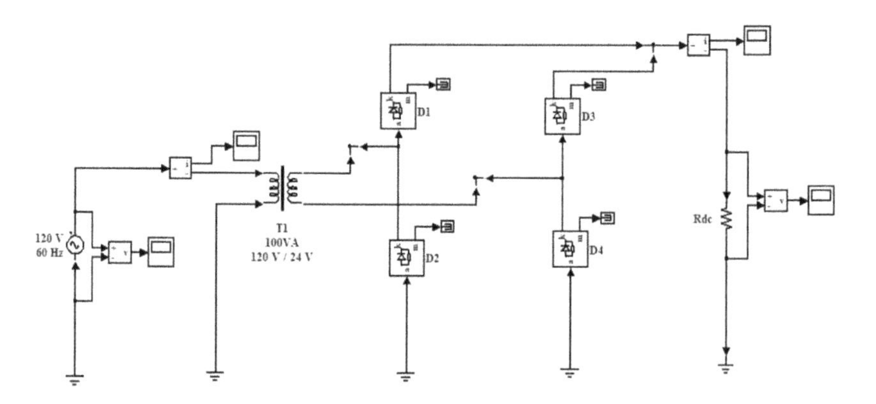

Figure 1.1 A typical single-phase AC-DC converter.

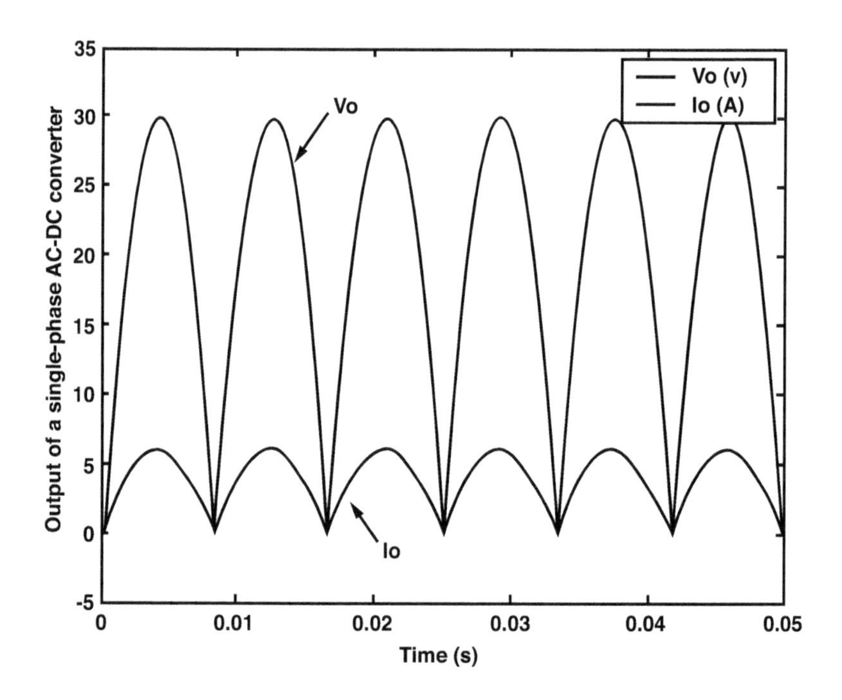

Figure 1.2 Output response of a single-phase AC-DC converter.

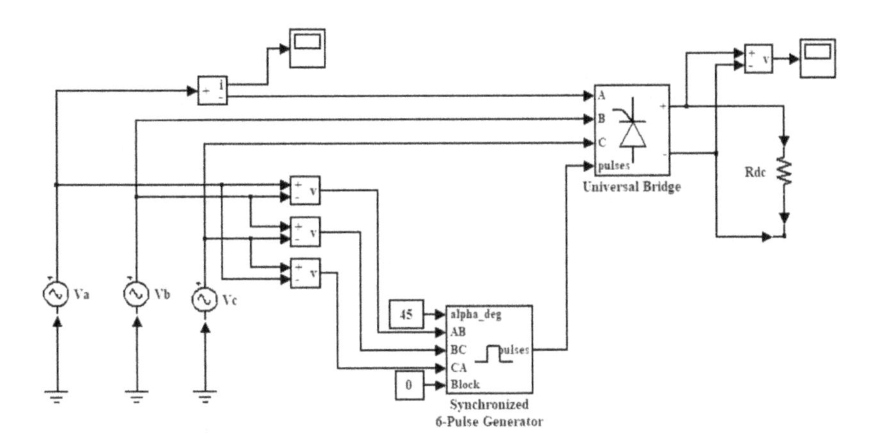

Figure 1.3 A three-phase AC-DC converter.

Table 1.2 System Specifications of the Three-Phase AC-DC Converter

Input			*Firing angle*	*Load resistance*
V_{ph} (V)	f_s (Hz)	Phase sequence	α (°)	R_{dc} (Ω)
120	60	ABC	45	1

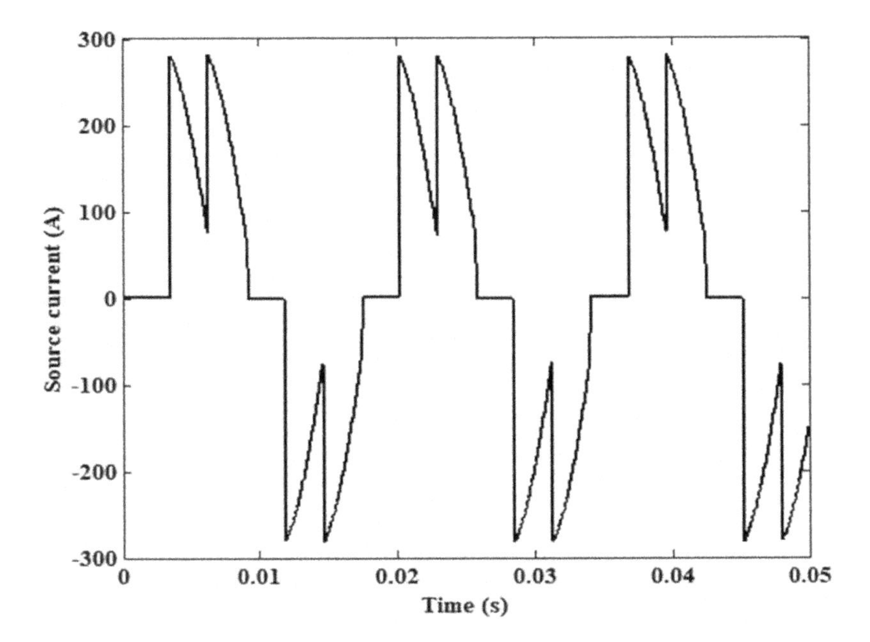

Figure 1.4 Source current of a three-phase AC-DC converter.

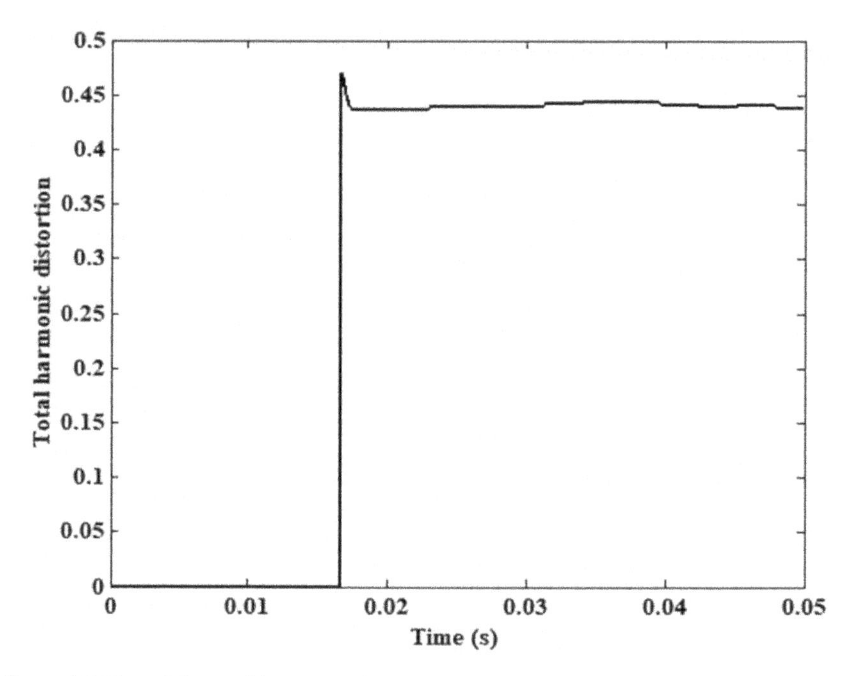

Figure 1.5 Plot of the total harmonic distortion.

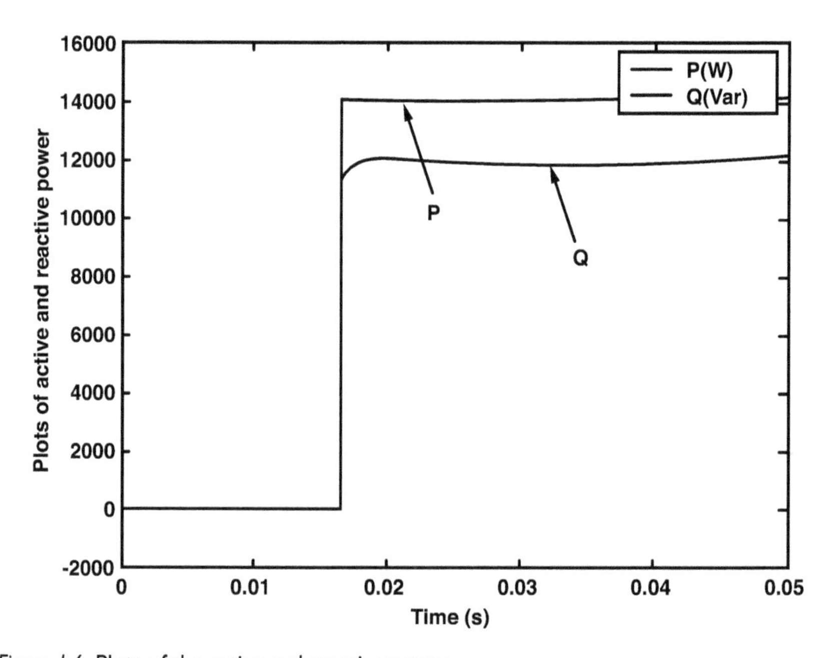

Figure 1.6 Plots of the active and reactive power.

1.3 MODELLING AND DIGITAL SIMULATION

Modelling and simulation play an important role in the scientific world in the 21st century. Simulation is technically used to oversee the practical scenario. Failure of components could be eliminated once a system is correctly modelled and satisfactory performance has been achieved from the simulation model. This would eventually reduce the time and cost involved in executing the laboratory prototype. MathWorks matrix laboratory (MATLAB)® is a scientific software package which has millions of users across industrial, academic, and research fields. Programming capability and availability of custom-made toolboxes enable MATLAB to interact with interdisciplinary areas of technology.

1.3.1 Sample control circuit using fuzzy logic controller

For example, as shown in Figure 1.7, continuous/discrete/math models available in the Simulink library of the MATLAB can be interconnected with other toolboxes like the Fuzzy Logic Toolbox to study the performance of an active power filter (APF) that includes the models of sources, static power converters, controllers, and sinks.

1.3.2 Modelling using simulink

One may develop the model and conduct simulation of a simple system first to grasp the basic background in modelling and simulation. Consider a system to determine the absolute value of a number; to start with modelling one may open a new model file from the MATLAB window and start dragging

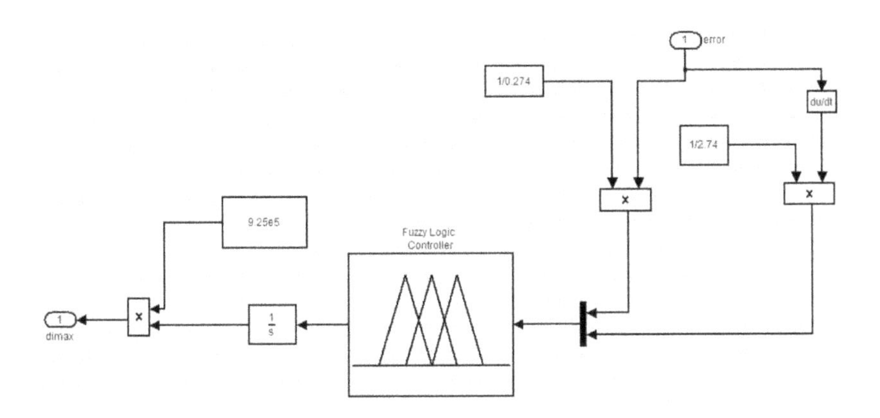

Figure 1.7 A Sample control circuit of an APF system using fuzzy logic controller.

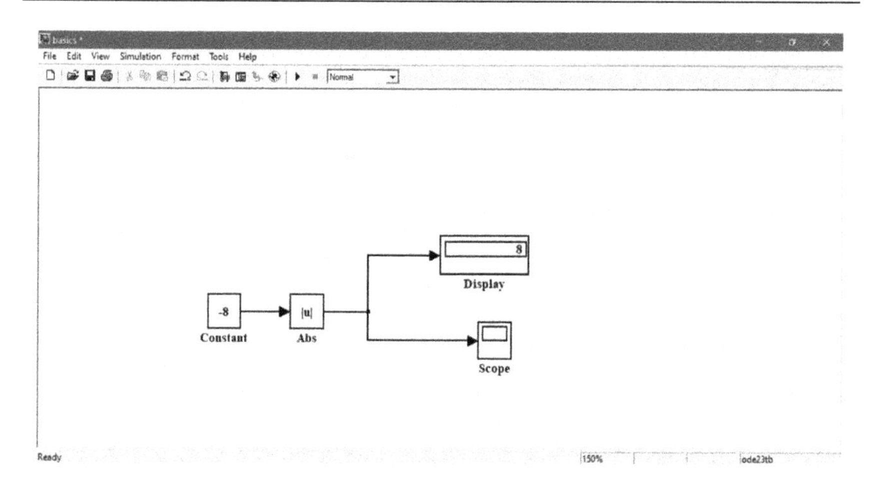

Figure 1.8 Simulink model to determine the absolute value of a number.

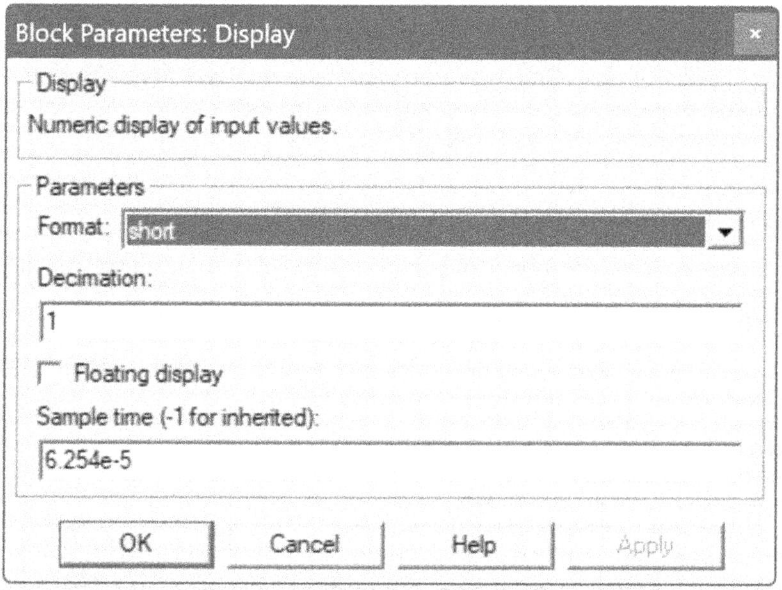

Block Parameters: Display
Display
Numeric display of input values.
Parameters
Format: short
Decimation:
1
Floating display
Sample time (-1 for inherited):
6.254e-5
OK Cancel Help Apply

Figure 1.9 Parameter settings of the display block.

Constant, Abs, Display and Scope blocks into the model file and connect the blocks as shown in Figure 1.8. One may choose the format of the Display block short, a default decimation of one to display data at every time step, a sample time of $6.254e^{-5}$ s to process the data at regular intervals as shown in Figure 1.9. In the general setting of the Scope block, the number of axes is chosen as "1" to display the output only, as shown in Figure 1.10. The

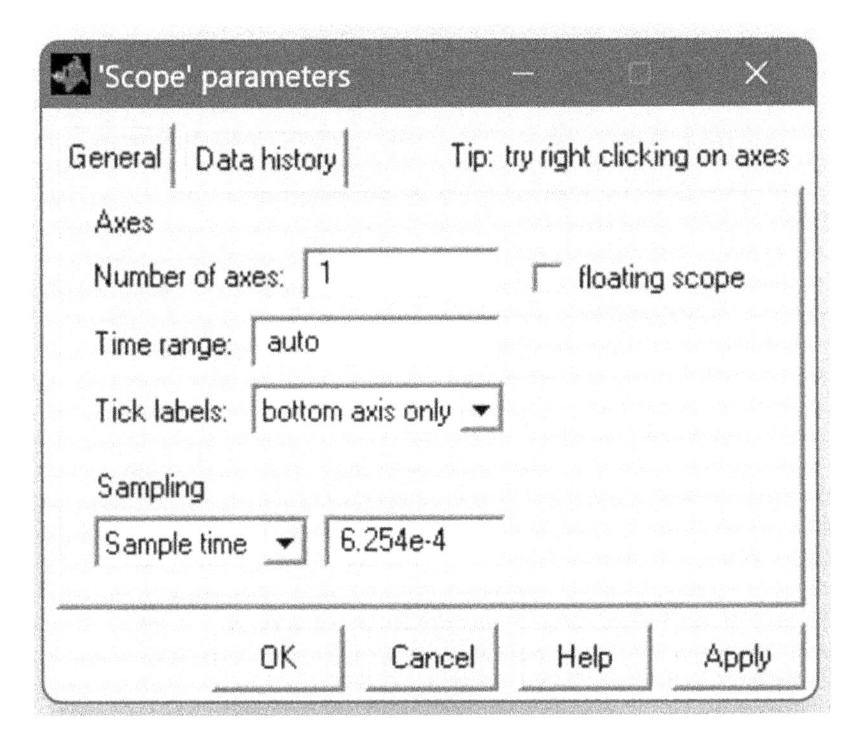

Figure 1.10 General parameter settings of the scope block.

number of axes can be more than 1 if the data is not saving in the workspace. To save data to the workspace one may choose a variable name with array format as shown in Figure 1.11. Simulation time and solver options are chosen as shown in Figure 1.12 and default settings are used for the rest of the parameters. To plot the output as a figure file as shown in Figure 1.13, one may use the following MATLAB commands:

```
>> save -ascii absvalue absvalue
  >> plot(absvalue(:,1),absvalue(:,2))
```

Tool bar options shown in Figure 1.13 could be used to format the figure file.

1.4 MODELLING OF A DIODE AS A SWITCH

MATLAB/Simulink/Power System Blockset (PSB) provides an effective platform to model electrical systems. PSB is a graphical tool that allows modelling and simulation of static power converters, generation, transmission, distribution, and consumption of electrical power in Simulink environment.

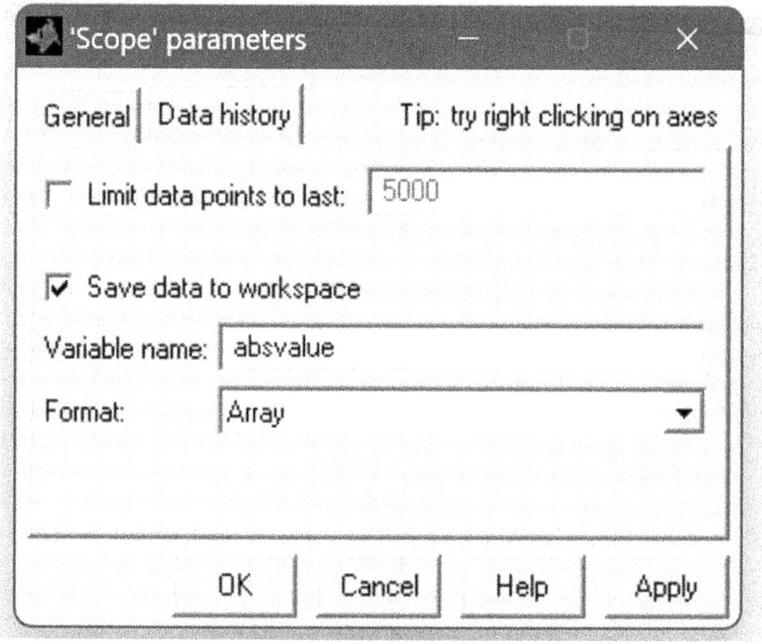

'Scope' parameters

General | Data history | Tip: try right clicking on axes

Limit data points to last: 5000

Save data to workspace

Variable name: absvalue

Format: Array

OK | Cancel | Help | Apply

Figure 1.11 Data history settings of the scope block.

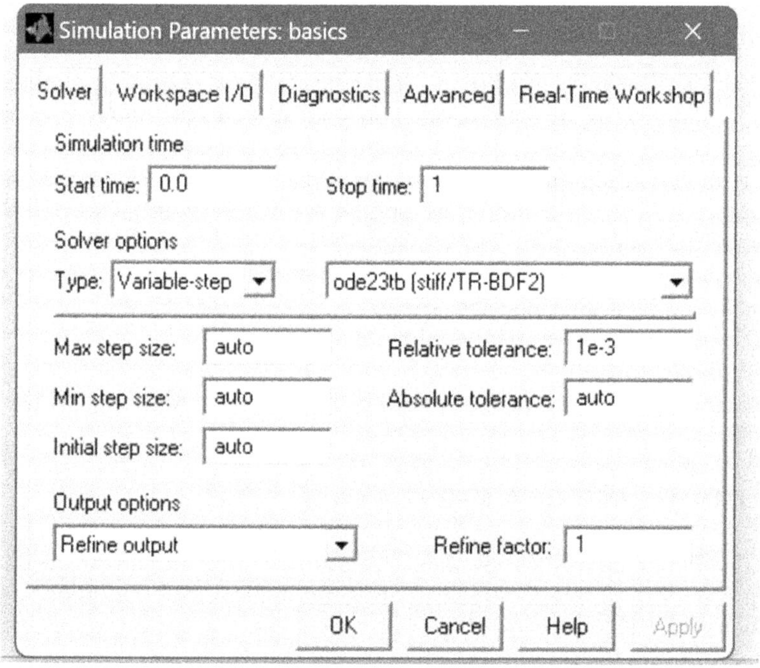

Simulation Parameters: basics

Solver | Workspace I/O | Diagnostics | Advanced | Real-Time Workshop

Simulation time

Start time: 0.0 Stop time: 1

Solver options

Type: Variable-step ode23tb (stiff/TR-BDF2)

Max step size: auto Relative tolerance: 1e-3

Min step size: auto Absolute tolerance: auto

Initial step size: auto

Output options

Refine output Refine factor: 1

OK | Cancel | Help | Apply

Figure 1.12 Simulation parameter settings.

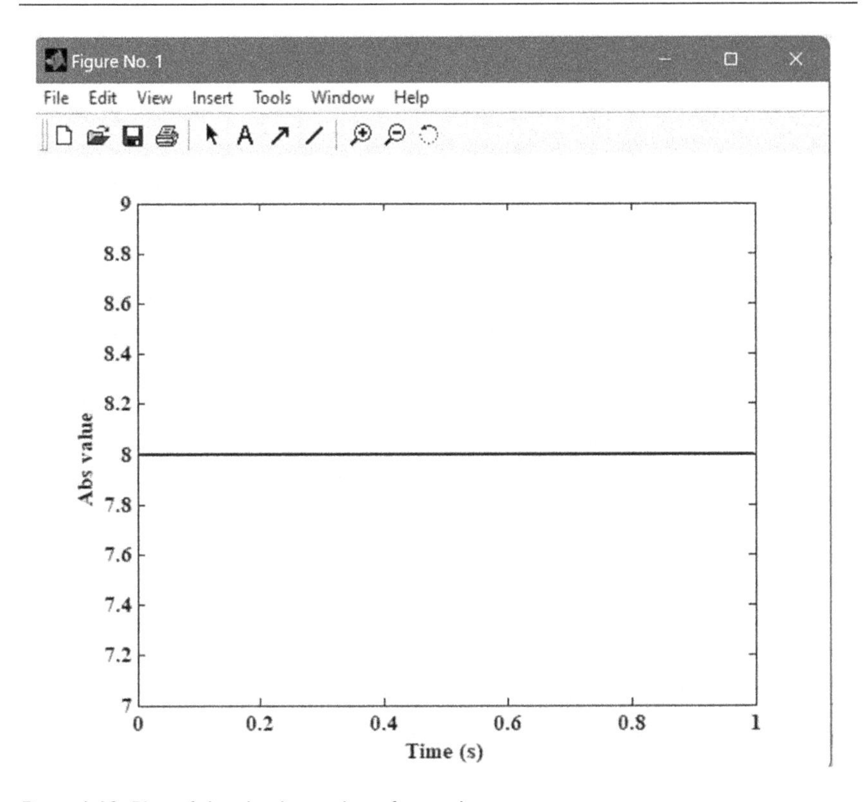

Figure 1.13 Plot of the absolute value of a number.

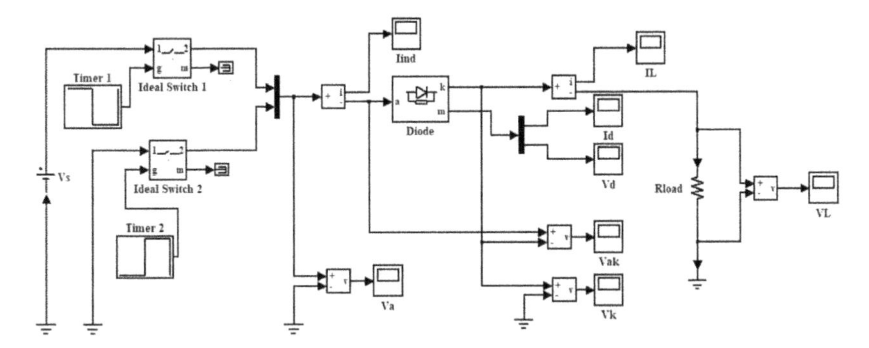

Figure 1.14 Diode as a switch in MATLAB/Simulink/PSB.

In Figure 1.14, a diode is used as a switch to connect input power supply to the load with system specifications as seen in Table 1.3. By configuring the ideal switches, a positive voltage V_s and the ground are alternatively connected to the anode of the diode. When V_s is connected to the anode, the diode would be in the ON position and connects the input V_s to the load; alternatively, when the ground is connected to the anode, the diode would be in the OFF position and disconnects load from the input V_s.

Table 1.3 System specifications of modelling diode as a switch.

Input		*ctimer₁*		*ctimer₂*		*Load resistance*
V_s (V)	V_e (V)	ON time (s)	OFF time (s)	ON time (s)	OFF time (s)	R_{load} (Ω)
9	0	0–0.05 and 0.1–0.15	0.05–0.1	0.05–0.1	0–0.05 and 0.1–0.15	100

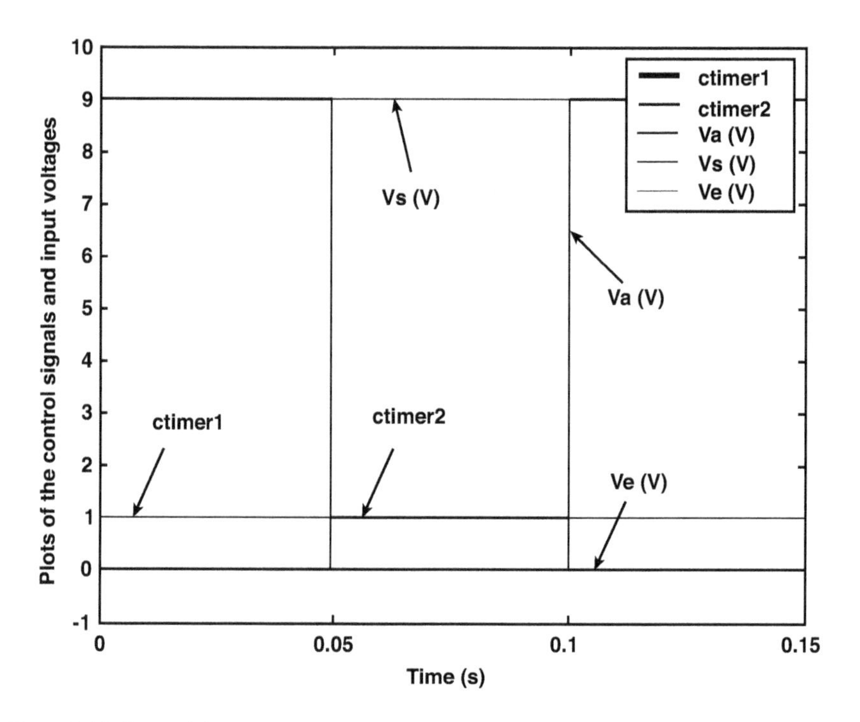

Figure 1.15 Plots of the timer outputs and input voltages.

As shown in Figure 1.14, the input V_s is connected to the anode of the diode through ideal switch$_1$ when the control signal of Timer$_1$ (ctimer$_1$) is 1; and the anode is connected to the ground through ideal switch$_2$ when the control signal of Timer$_2$ (ctimer$_2$) is 1. Analysis of Figure 1.15 shows that from 0 to 0.05 s, and 0.1 s to 0.15 s, ctimer$_1$ is 1 and V_a is equal to V_s. From 0.05 s to 0.1 s ctimer$_2$ is 1 and V_a is equal to V_e. As shown in Figure 1.16,

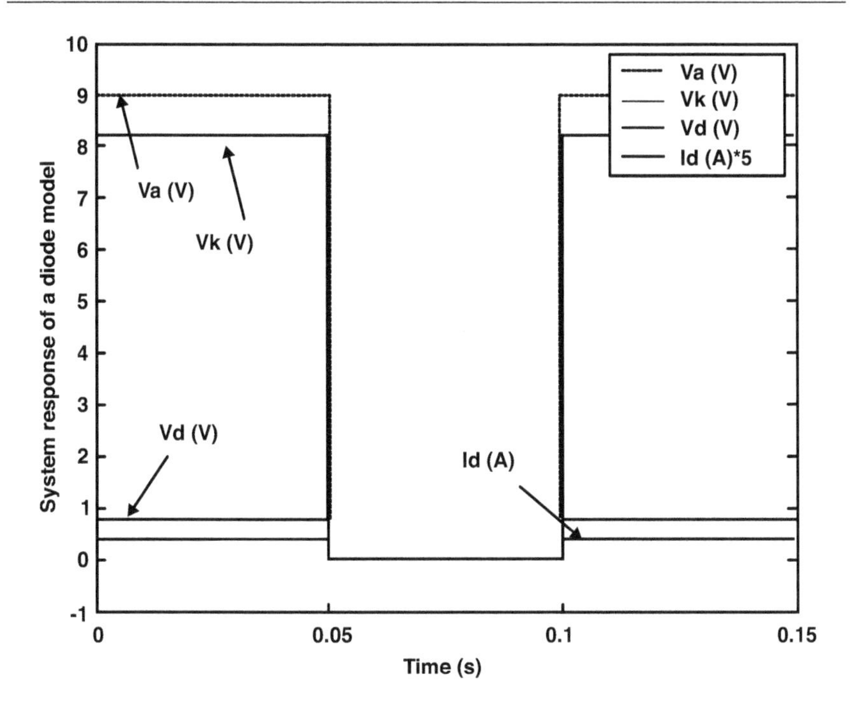

Figure 1.16 Plots of the system response of a diode model.

when V_s is connected to the anode, the diode conducts and current I_d flows from V_s to the load resistance $R_{load} = 100\ \Omega$; one may note that when diode conducts the voltage $V_d = V_{ak}$ across the diode is 0.8 V. Alternatively when the anode is connected to the ground, diode stops conducting and the load current $I_L = I_d$ is zero.

1.5 MODELLING OF A MOSFET AS A SWITCH

In Figure 1.17, a MOSFET is configured as a fully controllable switch with system specifications as seen in Table 1.4 below. Analysis of Figure 1.18 shows that, with the circuit configuration as shown in Figure 1.17, ON/OFF

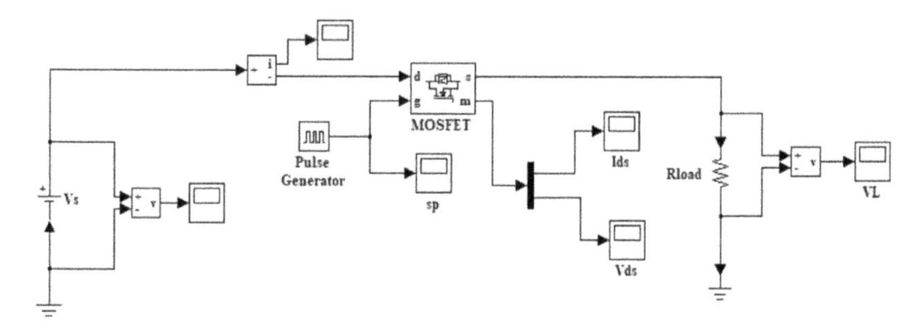

Figure 1.17 MOSFET as a fully controllable switch in MATLAB/Simulink/PSB.

Table 1.4 System specifications of modelling MOSFET as a switch.

Input	*Switching pulses*			*Load resistance*
V_s (V)	Period T (s)	Duty cycle D	Amplitude (V)	R_{load} (Ω)
9	1 e^{-3}	50 %	1	5

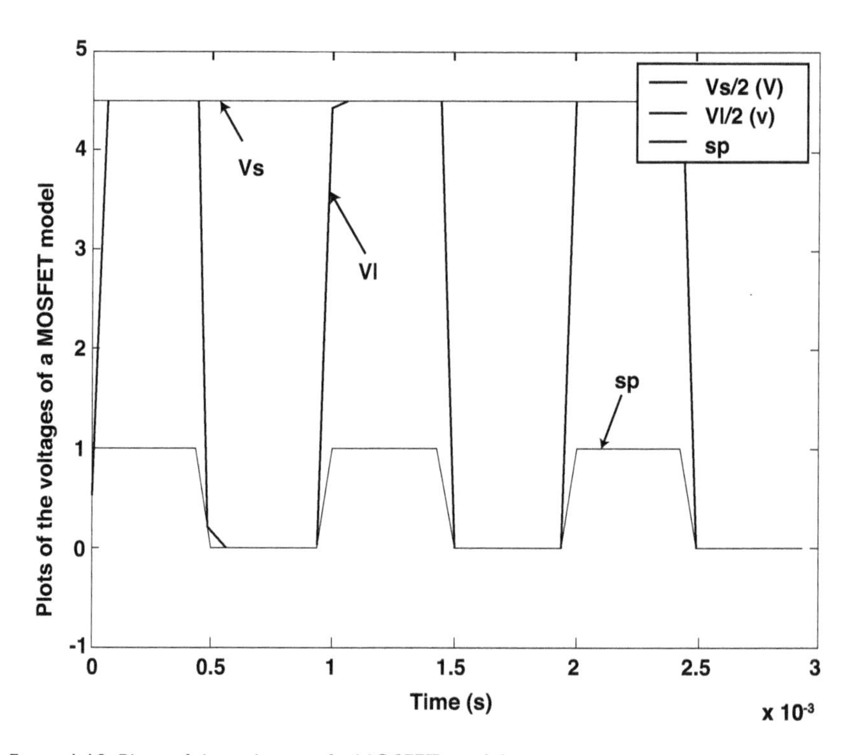

Figure 1.18 Plots of the voltages of a MOSFET model.

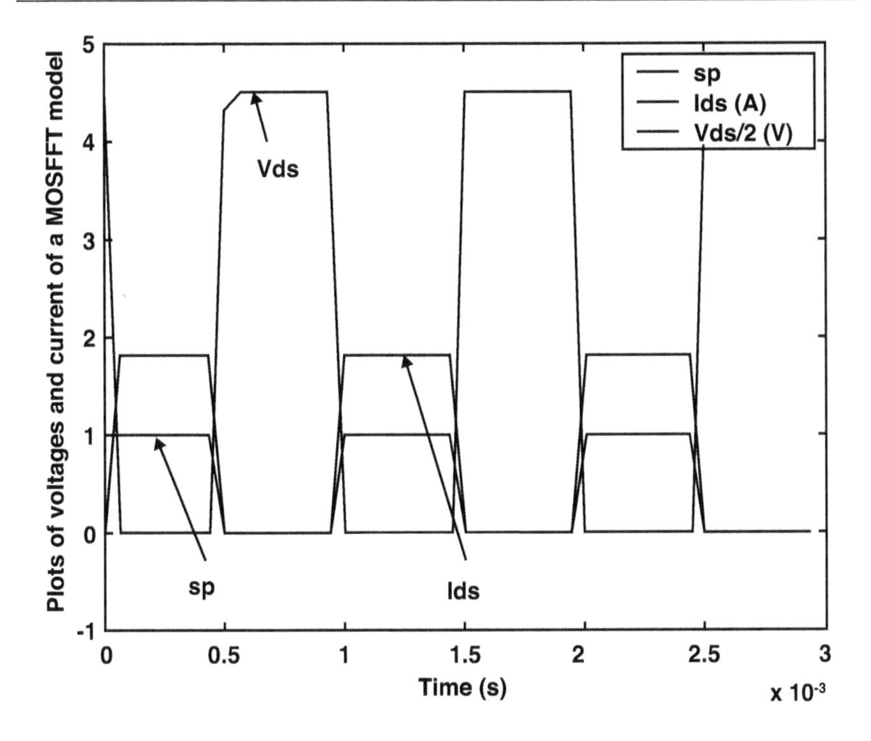

Figure 1.19 Plots of the system response of a MOSFET model.

switching of the MOSFET is fully controlled by the switching pulses (sp) from the pulse generator.

1.6 PSB MODEL OF A BUCK CONVERTER

Figure 1.20 shows the PSB model of a buck converter. As shown in Figure 1.20, unlike the heavy weight transformer-based AC-DC circuits, the power circuit of the buck converter consists of just two energy storage devices and two switching elements making it suitable for various applications like USB on-the-go, solar chargers, quadcopters etc.

The inductor, one of the energy storage elements of the buck converter, stores kinetic energy and releases the stored energy by reversing its voltage polarity; the capacitor, the other energy storage element, stores potential energy and releases the stored energy by reversing the polarity of the current through it. The back and forth switching of semiconductor switches controls the charging and discharging patterns of these storage elements. The turn on

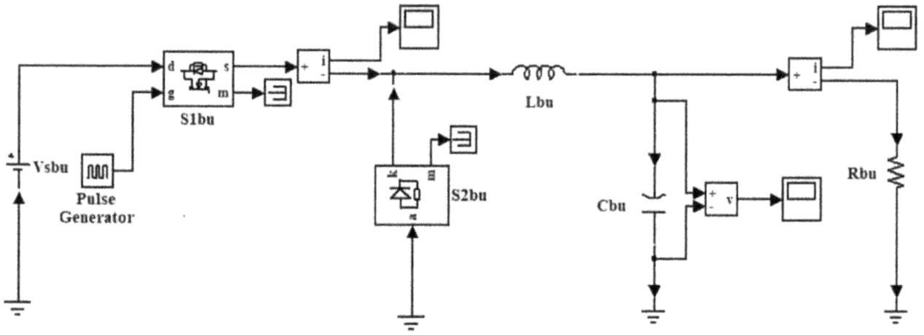

Figure 1.20 PSB model of a buck converter.

and turn off operation of these switches are determined by control circuits designed for specific application.

BIBLIOGRAPHY

Akagi H., Trends in active power line conditioners. *IEEE Transactions on Power Electronics*, 9(3), 263–268, 1994.

Ashok S. and Preetha P., Power quality and custom power under deregulated power system. India: Proceedings of the ISTE-AICTE Short Term Training Program, 2004.

Bimbra P. S., Power electronics (4th ed.). India: Khanna Publishers, 2006.

Bose B. K., Modern power electronics and AC drives (1st ed.). India: Pearson Education Inc., 2002.

Bradley D. A., Power electronics (2nd ed.). England: Chapman & Hall, 1995.

Driankov D., Hellendoorn H., and Reinfrank M., An introduction to fuzzy control (student ed.). India: Springer International, 1993.

George M., Advanced active power line conditioners for power quality improvement. Malaysia: Multimedia University, 2009.

George M., Basu K. P., and Younis M. A. A., Digital simulation of static power converters using power system blockset (1st ed.). Germany: LAP LAMBERT Academic Publishing, 2012.

Hadi S., Power system analysis (2nd ed.). Singapore: McGraw-Hill Education (Asia), 2004.

Hughes E., Electrical and electronics technology (10th ed.). UK: Pearson Education Limited, 2008.

Klir G. J. and Yuan B., Fuzzy sets and fuzzy logic theory and applications (eastern economy edition, 7th reprint). India: Prentice-Hall of India Private Limited, 2002.

MATLAB Education Seminar. Malaysia: International Islamic University, 2003.

Mohan N., Power electronics modeling simplified using PSPICE (Release 9). Canada: University of Minnesota, 2002.

Mohan N., Undeland T. M., and Robbins W. P., Power electronics: converters, applications and design (3rd ed.). USA: John Wiley & Sons Inc., 2003.
Part-Enander E., Sjoberg A., Melin B., and Ishaksson P., The MATLAB handbook (1st ed.). USA: Addison Wesley, 1996.
Rai H. C., Industrial and power electronics (1st ed.). India: Umesh Publications, 1987.
Rashid M. H., 2 days course on power electronics and its applications. Malaysia: Universiti Putra Malaysia, 2004.
Rashid M. H., Power electronics circuits, devices and applications (3rd ed.), India: Pearson education Inc., 2004.
Rashid M. H., Tutorial on design and analysis of power converters. Malaysia: Universiti Putra Malaysia, 2002.
Subramanyam V., Power electronics (1st ed.). India: New Age International Private Limited, 2003.
The MathWorks Inc., Power system blockset for use with Simulink. USA: The MathWorks Inc., 2000.
Wildi T., Electrical machines, drives, and power systems (4th ed.). USA: Prentice Hall International Inc., 2000.

Chapter 2

Switching characteristics of energy storage elements

2.1 INTRODUCTION

One may use the step response to illustrate the switching characteristics of the energy storing elements when the input changes from one value to another value in a very short time. Unlike a resistance (R), the voltage across a capacitor (C) and the current through an inductor won't change instantaneously. Also, the switching behaviour of energy storage elements depends on the properties of these elements as well as on the circuit resistance. Resistance of the electrical load should be taken into consideration when the system behaviour is analyzed. Thus, it is necessary to study the behaviour of the energy storage elements for different parameter values of the circuit elements.

2.2 ENERGY STORAGE IN AN R-C CIRCUIT

Consider an R-C circuit, as shown in Figure 2.1, configured to operate in two different modes; in mode 1 the capacitor stores energy from the DC source, and in mode 2 the stored energy of the capacitor will be delivered to the load by controlling the turn ON and turn OFF time of the switches SW_1 and SW_2. One may note that ON/OFF times of switches SW_1 and SW_2 are complementary to each other.

Analysis of Figure 2.2 shows that in mode 1, the switch SW_1 is in the ON position connecting the source to the load and the switch SW_2 is in the OFF position disconnecting the load from the capacitor. Analysis of Figure 2.3 shows that in mode 2, the SW_1 would be OFF and SW_2 would be ON, disconnecting the source and connecting the load to the capacitor. Timers are set for equal time setting, but complementary operation, for a total duration of 0.14 s. Analysis of Figure 2.4, Figure 2.5, and Figure 2.6 shows that from 0–0.07 s, the circuit operates in mode 1 and from 0.07 s–0.14 s the circuit operates in mode 2. From 0–0.07 s the voltage across the capacitor increases and from 0.07 s–0.14 s the voltage across the capacitor decreases in an

DOI: 10.1201/9781003511236-2

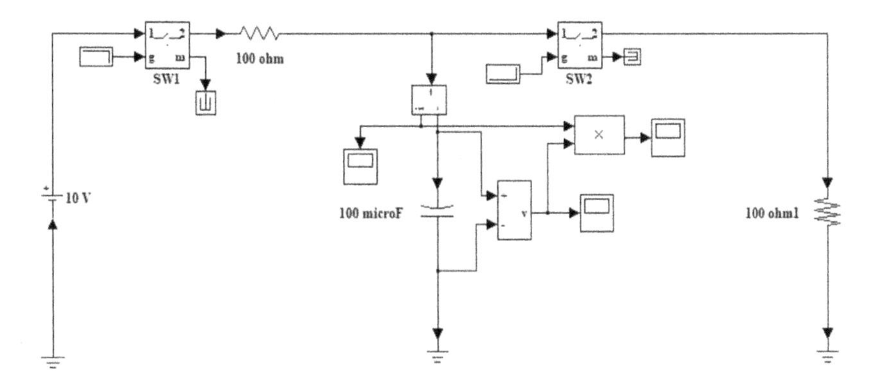

Figure 2.1 Configuration of a capacitive circuit to store/deliver energy.

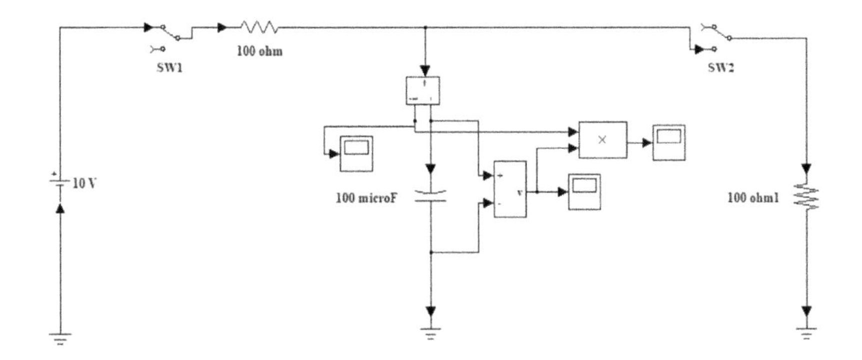

Figure 2.2 Configuration of a capacitive circuit in mode 1 to store energy.

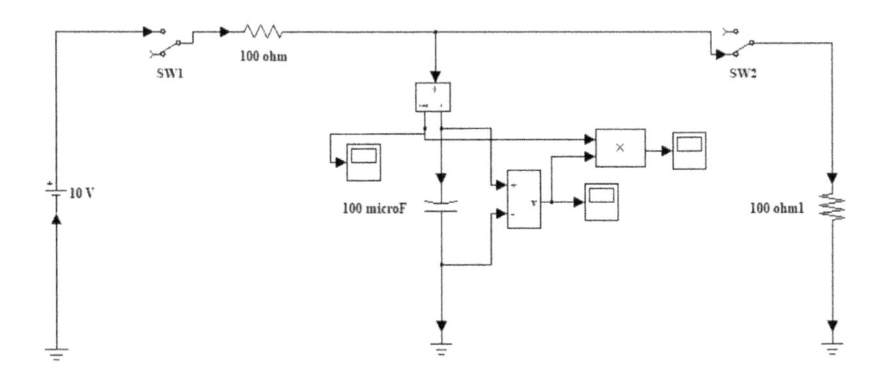

Figure 2.3 Configuration of a capacitive circuit in mode 2 delivering energy.

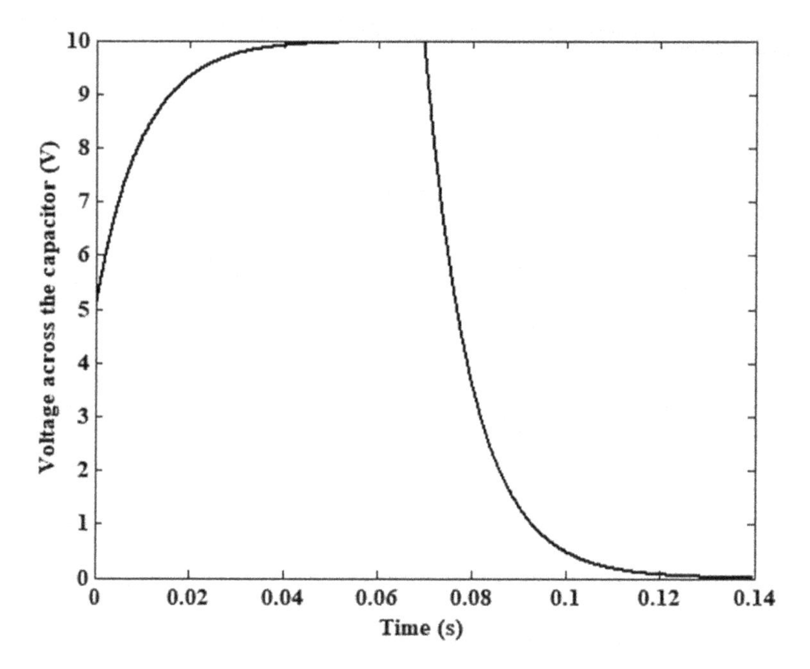

Figure 2.4 Plot of instantaneous values of voltage across the capacitor.

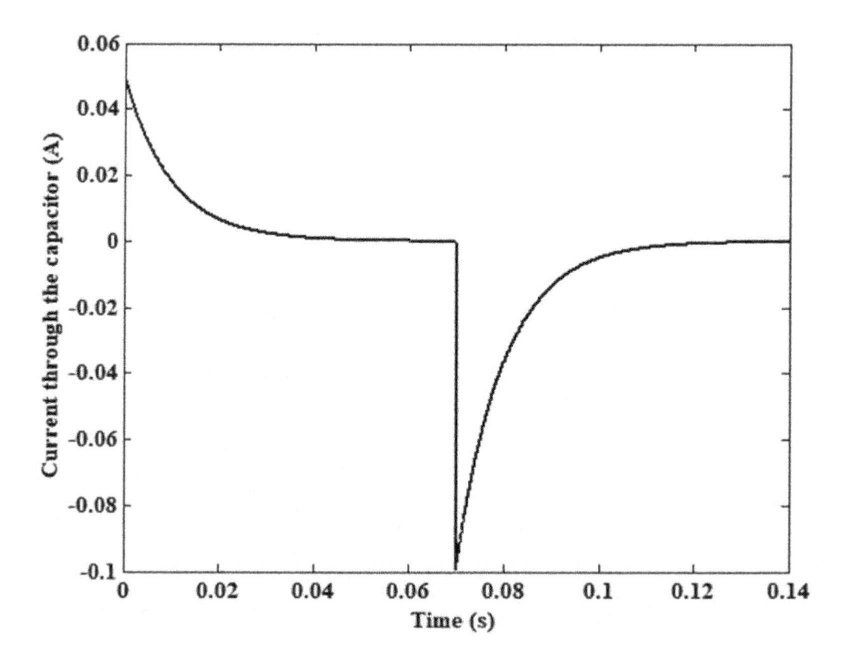

Figure 2.5 Plot of instantaneous values of current through the capacitor.

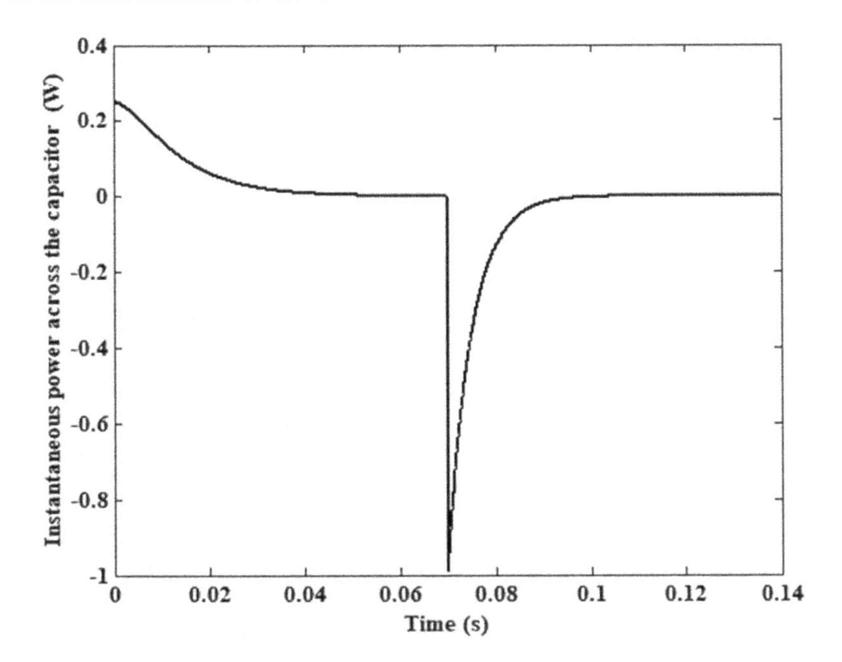

Figure 2.6 Plot of instantaneous power across the capacitor.

exponential manner but the capacitor voltage never goes negative. Analysis of Figure 2.5 shows that as the capacitor builds up voltage from 0–0.07 s, the current through the capacitor decreases and when the capacitor is connected to the load from 0.07 s–0.14 s, the capacitor supply current to the load and the capacitor current becomes negative. Analysis of Figure 2.6 shows that in mode 1, the capacitor stores energy from the source and in mode 2, the stored energy would be delivered to the load; one may note that the area under instantaneous power p_c (W) vs time t (s) graph gives the energy associated with the capacitor; and p_c (W) is given by Eqn. (2.1),

$$p_c = v_c \times i_c \tag{2.1}$$

where v_c – instantaneous value of voltage across the capacitor (V); i_c – instantaneous value of current through the capacitor (A).

2.2.1 Switching characteristics of a capacitor with variations in switching frequency

One may start with the conventional approach to have a clear understanding of the circuit response for different values of resistance and capacitance. The time constant τ (RC), the time taken to reach 63.2% of the final steady state value while charging a capacitor, predicts how quickly the capacitor may

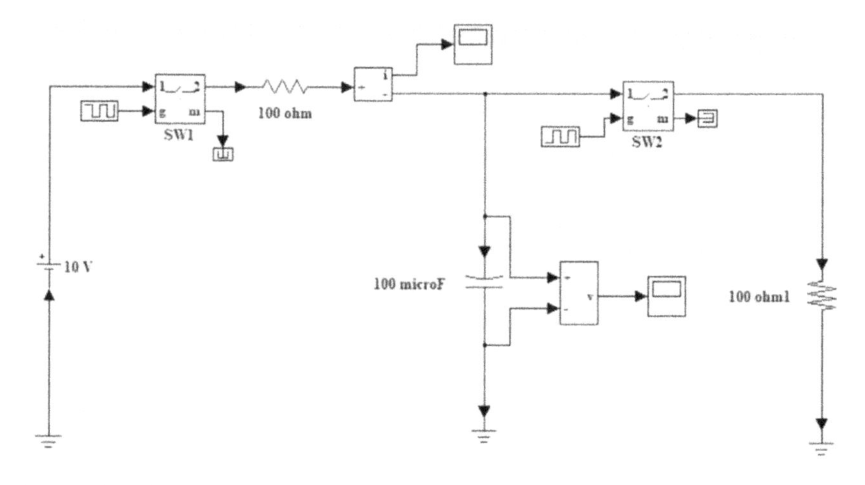

Figure 2.7 An R-C switching circuit.

charge. While designing the circuit, the capacitor should be given enough time to reach the steady state value; consider the R-C circuit shown in Figure 2.7, with R = 100 Ω; C = 100μF connected to a DC voltage source of 10 V; one may note that the time constant $\tau = RC = 0.01$s. The capacitor reaches approximate final steady state value when the time equals 5 τ.

Considering switch SW_1 turn on time, $T_{ONsw1} = 0.07$ s, with a 50% duty cycle, the time period of switch SW_1, $T_{sw1} = 0.14$ s and the corresponding switching frequency would be approximately equal to 7 Hz; one may also note that ON/OFF time of switch SW_2 has been set as complementary to that of SW_1. Analysis of Figure 2.8 shows that capacitor is getting sufficient time to reach the final steady state 10 V while charging and 0 V while discharging; however, if the switching frequency is increased, the capacitor may not get sufficient time to reach the final steady state value as illustrated in Figure 2.9 and Figure 2.10.

2.2.2 PSB model of an R-C charging circuit

The simulation model of a simple R-C charging circuit connected to the DC source available in the library of MATLAB®/Simulink/Power System Blockset is shown in Figure 2.11. To analyze the unit step response, the amplitude of the DC voltage source has been set to one volt; the series RLC branch available in the library of MATLAB/Simulink/Power System Blockset has been configured as resistive and capacitive elements as shown in Figure 2.11. The powergui block included in Figure 2.11 is needed to simulate models containing MATLAB/Simulink/Power System Blockset to store the equivalent state-space equations for the Simulink circuit.

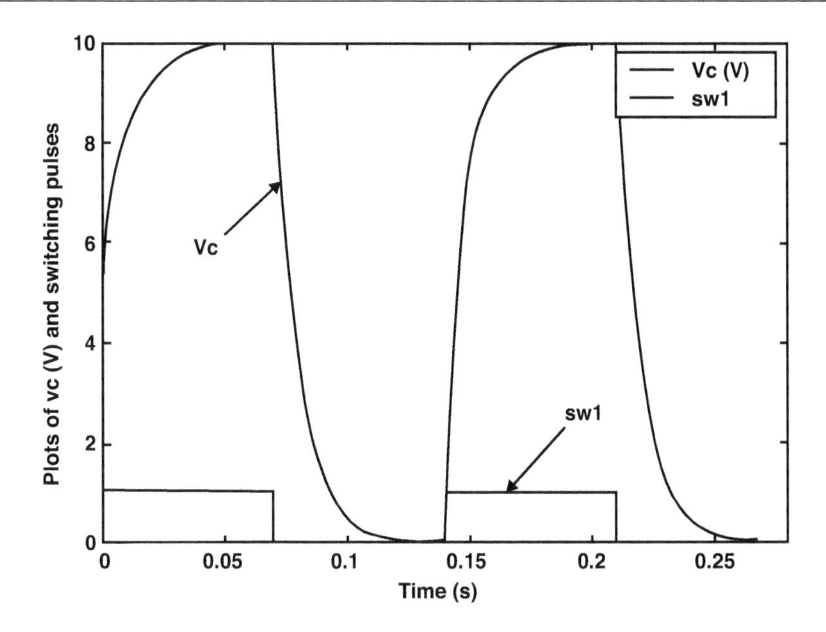

Figure 2.8 Illustration of the capacitor voltage and switching pulses for a frequency of 7 Hz.

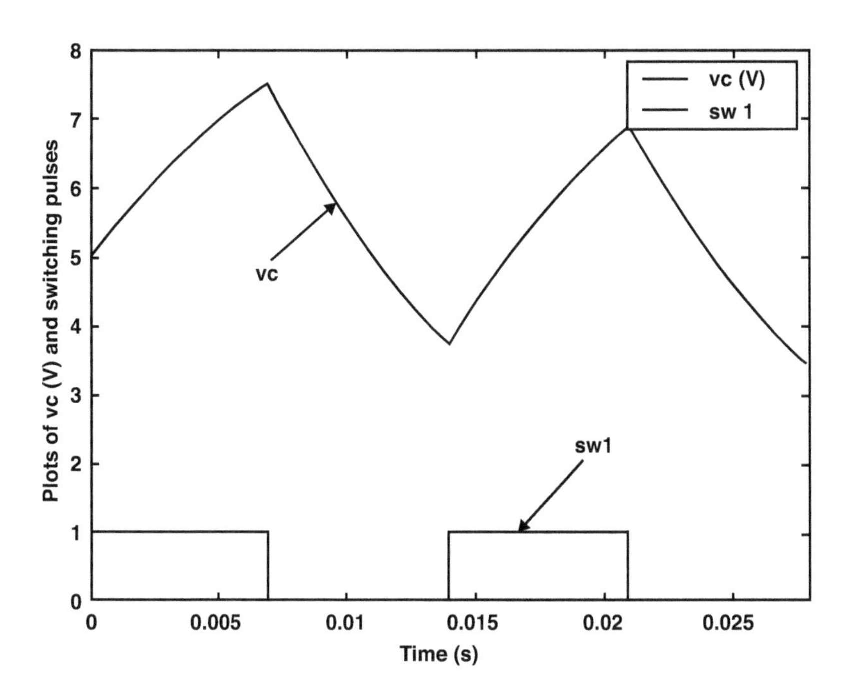

Figure 2.9 Illustration of the capacitor voltage and switching pulses for a frequency of 71 Hz.

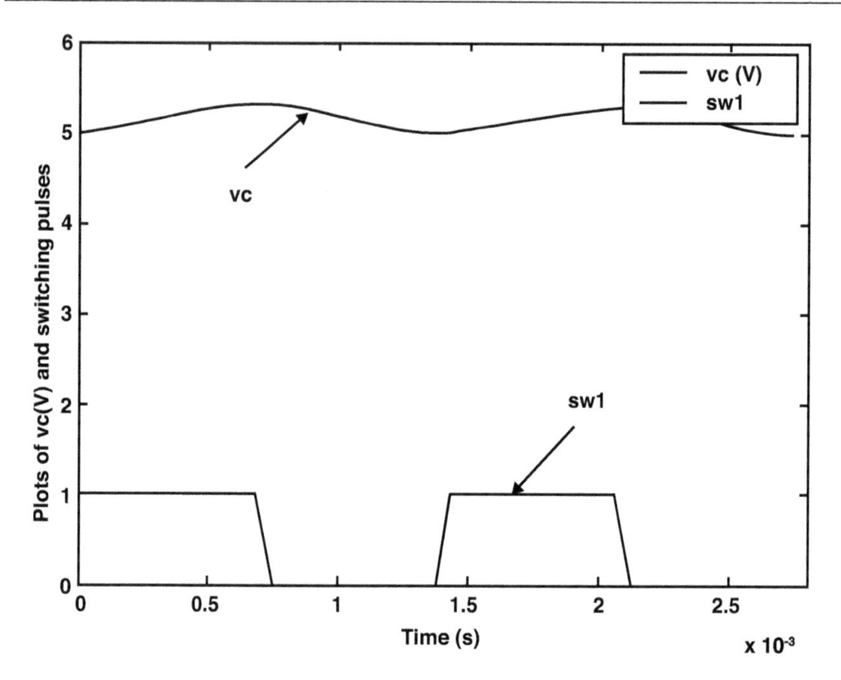

Figure 2.10 Illustration of the capacitor voltage and switching pulses for a frequency of 714 Hz.

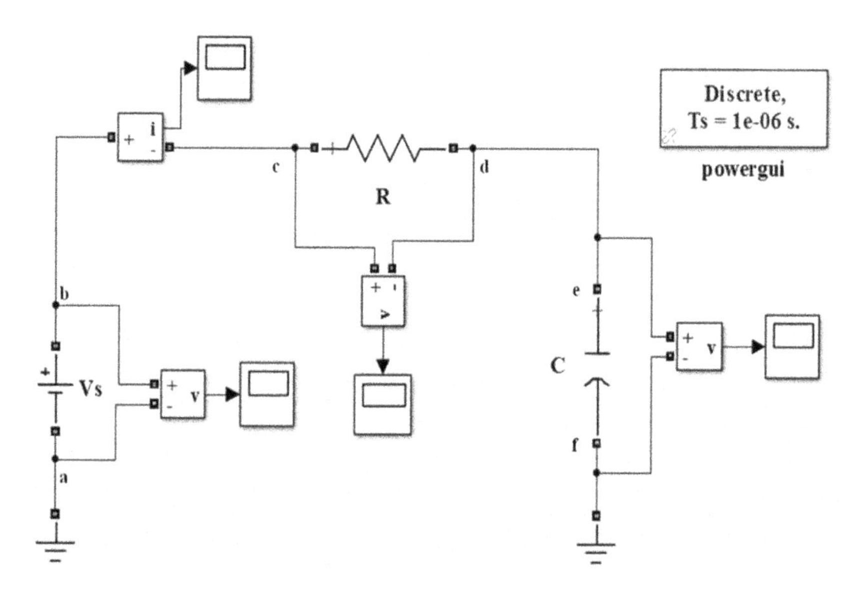

Figure 2.11 PSB model of an R-C circuit.

2.2.3 Mathematical formulation to determine the voltage across the capacitor

The voltage across the capacitor in Figure 2.11 varies with time and satisfies Eqn. (2.2)

$$v_c = V_s(1 - e^{-\frac{t}{\tau}})$$

where V_s—source voltage (V); v_c—instantaneous value of voltage across the capacitor (V); t—time (s); τ—time constant (s).

One may able to derive Eqn. (2.2) by applying Kirchoff's voltage law (KVL) to the circuit shown in Figure 2.11; consider the closed loop circuit **abcdefa** and assume that the loop current (i) flows in the clock-wise direction.

2.2.3.1 *Polarity of the voltage source*

As one moves from a lower potential **a** to a higher potential **b** there is a rise in voltage across the source voltage V_s and a rise in voltage should be treated as positive.

2.2.3.2 *Polarities of the voltage drop across circuit elements*

As far as the polarities of the voltage drop across the circuit elements are concerned, if the direction of movement is in the direction of current, as current flows from a higher potential to lower potential, there is a fall in potential and a fall in potential should be treated as negative.

Applying KVL to the closed loop circuit **abcdefa** of Figure 2.11, the algebraic sum of all the voltages in the closed loop **abcdefa** is equal to zero; and thus one may write Eqn. (2.3),

$$V_s - iR - v_c = 0$$

where i—instantaneous value of current through the network in the clock-wise direction (A); R—resistance (Ω).

Rewriting Eqn. (2.3), one may get,

$$V_s = iR + v_c$$

The current through the circuit is nothing but the current passing through the capacitor which is the differential of the charge (q) across the capacitor with respect to time. Thus Eqn. (2.4) could be written as:

$$V_s = \frac{d}{dt}(q) \times R + v_c$$

Substituting the expression of $q = Cv_c$ in Eqn. (2.5), one may get,

$$V_s = \frac{d}{dt}(Cv_c) \times R + v_c$$

where C—capacitance (F); rearranging Eqn. (2.6), one may get,

$$V_s - v_c = RC \times \frac{d}{dt}(v_C)$$

$$\frac{d(v_c)}{V_s - v_c} = \frac{dt}{RC}$$

Integrating Eqn. (2.8), one may get,

$$\log_e(V_s - v_c) = -\frac{t}{RC} + K$$

when $t = 0$; $v_c = 0$ which implies that $\log_e(V_s) = K$; thus,

$$\log_e(V_s - v_c) = -\frac{t}{RC} + \log_e(V_s)$$

$$\log_e\left(\frac{V_s - v_c}{V_s}\right) = -\frac{t}{RC}$$

or in other words,

$$\frac{V_s - v_c}{Vs} = e^{-\frac{t}{RC}}$$

$$\therefore v_c = V_s\left(1 - e^{-\frac{t}{\tau}}\right)$$

where RC is the time constant of the network during which capacitor voltage reaches 63.2% of the final steady state value while charging.

That's when $t = RC$

$$v_c = V_s\left(1 - e^{-RC/RC}\right) = V_s\left(1 - e^{-1}\right) = V_s\left(1 - \frac{1}{e}\right) = V_s\left(1 - \frac{1}{2.7183}\right) = 0.632V_s$$

2.2.4 Charging characteristics of the capacitor using conventional approach

Charging characteristics of the R-C circuit for different values of time constant has been obtained by varying values of resistance and capacitance in Figure 2.11 as seen in Table 2.1. Parameter settings (1) (para#1) has the least

Table 2.1 Voltage across the Capacitor with respect to Time for Different Values of time Constant

Time(s)	v_c (V)
para#1 R = 10 Ω; C = 10 μF; RC = 0.0001 s	
Time(s)	v_c (V)
0	0*
0.0001	0.632
0.0002	0.8647
0.0004	0.9817
0.0006	0.9975
0.001	1**
0.002	1
para#2 R = 10 Ω; C = 100 μF; RC = 0.001 s	
Time(s)	v_c (V)
0	0
0.001	0.632
0.002	0.8647
0.004	0.9817
0.006	0.9975
0.008	0.9997
0.01	1
para#3 R = 100 Ω; C = 100 μF; RC = 0.01 s	
Time(s)	v_c (V)
0	0
0.001	0.0952
0.002	0.1813
0.004	0.3297
0.006	0.4512
0.008	0.5507
0.01	0.632
0.02	0.8647
0.04	0.9817
0.06	0.9975
0.08	0.9997
0.1	1

*when $t = 0, v_c = 1\left(1 - e^{-0/0.0001}\right)$

**when $t = 0.001, 1\left(1 - e^{-0.001/0.0001}\right) = 1$

time constant as compared to second and third settings. The time taken to reach the steady state value depends on the time constant; as the time constant increases, the time taken to reach the steady state value also increases. Analysis of Figure 2.12 shows that with para#1, the capacitor voltage reaches the final steady state value faster than with para#2 and para#3.

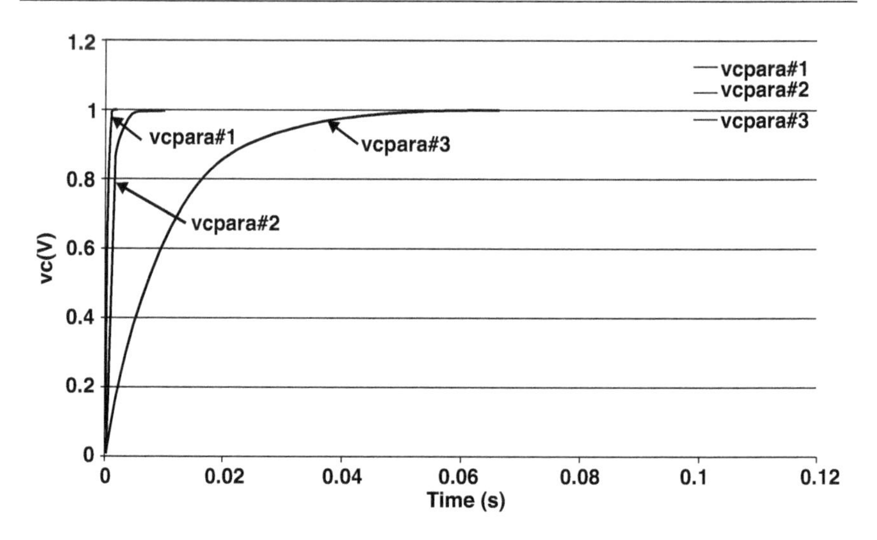

Figure 2.12 Charging characteristics of a capacitor using conventional approach.

2.2.5 Illustration of system response using PSB model

Figure 2.13 shows the charging characteristics of a capacitor with unit step input using the PSB model in Figure 2.11. One may compare the step response in Figure 2.13 with the charging characteristics plotted in Figure 2.12 to validate the employability of MATLAB/Simulink/PSB library to model electrical circuits. Analysis of the system response of para# 2 plotted in Figure 2.14 shows that as the capacitor charges, the voltage drop across the resistance decreases in an exponential manner and finally reaches zero and when the capacitor reaches final steady state value the current through the circuit is zero; meaning that magnitude of the capacitor voltage is the same as that of the source voltage but tends to drive a current in the opposite direction as that of the input voltage source.

2.3 ENERGY STORAGE IN AN R-L CIRCUIT

Consider an R-L circuit shown in Figure 2.15 configured to operate in two different modes; in mode 1 the inductor stores energy from the DC source and in mode 2 the stored energy of the inductor will be delivered to the load by controlling the turn ON and turn OFF time of the switch SW_1 (MOSFET). One may note that ON/OFF times of switches SW_1 and SW_2 are complementary to each other.

The circuit configuration shown in Figure 2.16 shows that in mode 1 the switch SW_1 would be turned ON by the gate input (g); one may note that when SW_1 conducts, the diode (SW_2) is reverse biased and would be in the OFF position.

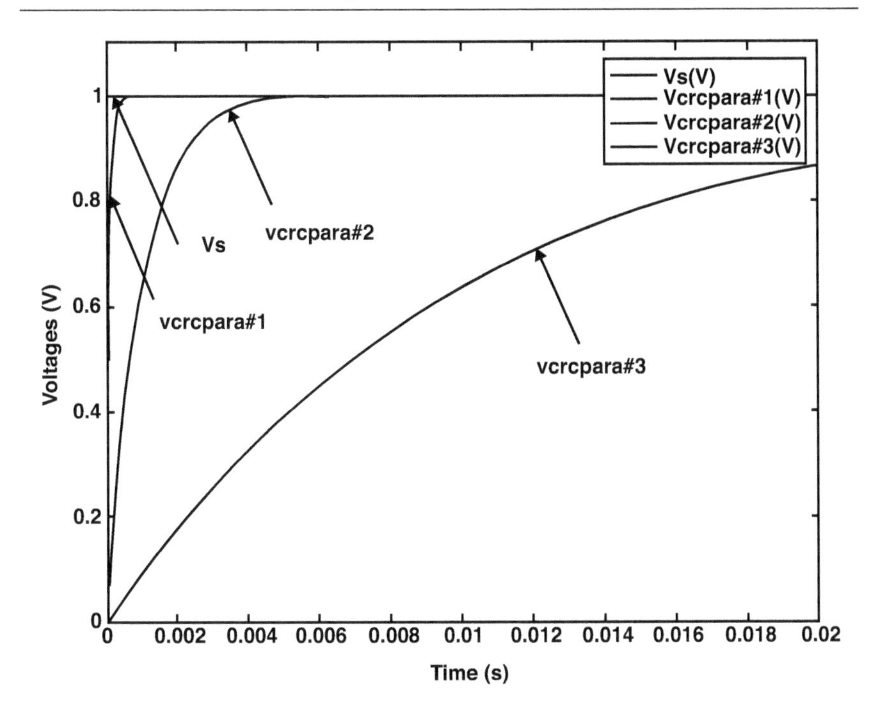

Figure 2.13 Step response of an R-C circuit using PSB model.

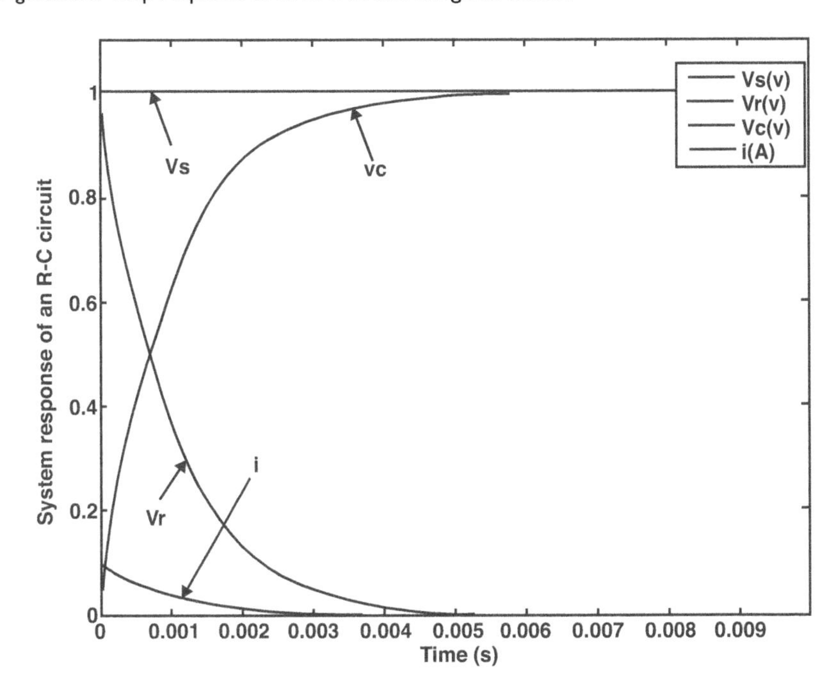

Figure 2.14 System response of an R-C circuit using PSB model.

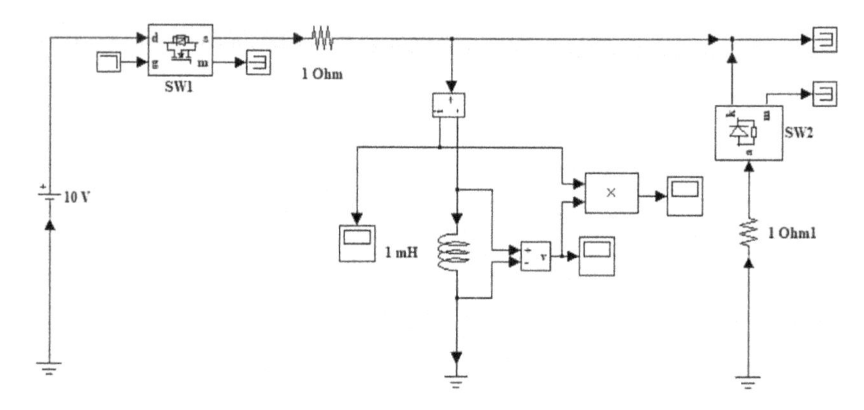

Figure 2.15 Configuration of an inductive circuit to store/deliver energy.

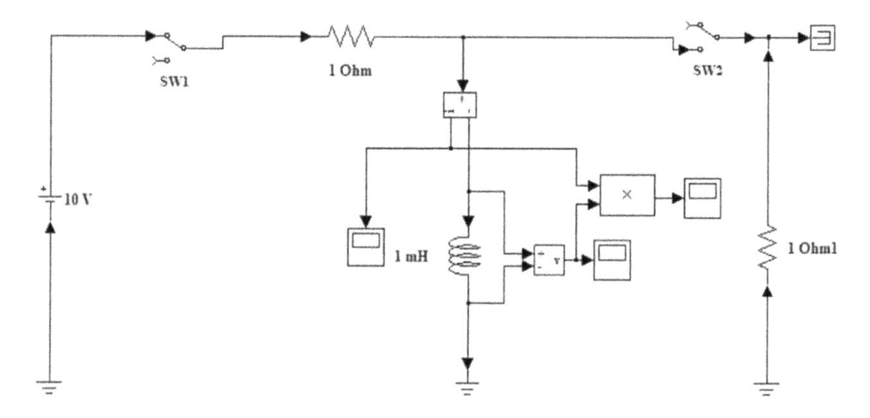

Figure 2.16 Configuration of an inductive circuit in mode 1 to store energy.

The circuit configuration shown in Figure 2.17 shows that in mode 2 the switch SW_1 would be turned OFF by the gate input (g) and the diode (SW_2) would be in the ON position.

Analysis of Figure 2.18 and Figure 2.19 shows that in mode 1 from 0–0.007 s as the current through the inductor increases exponentially, the voltage across the inductor decreases in an exponential manner. In mode 2, from 0.007 s–0.012 s, as soon as switch SW_1 is in the OFF position voltage across the inductor reverses polarity as shown in Figure 2.19 and forward biases the diode; one may note that in mode 2, the inductor delivers energy to the load. Analysis of Figure 2.18 shows that in mode 2, the inductor current starts decreasing and reaches zero but never goes negative.

Analysis of Figure 2.20 shows that in mode 1, the inductor stores energy from the source and in mode 2, the stored energy would be delivered to the load; one may note that the area under instantaneous power p_l (W) vs time

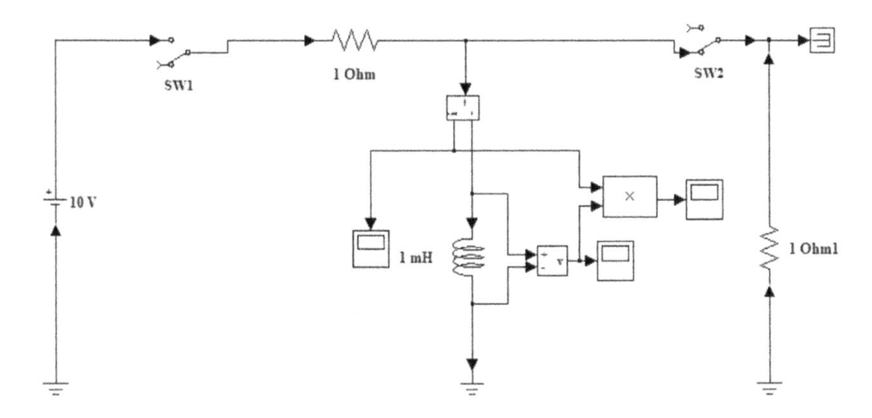

Figure 2.17 Configuration of an inductive circuit in mode 2 delivering energy.

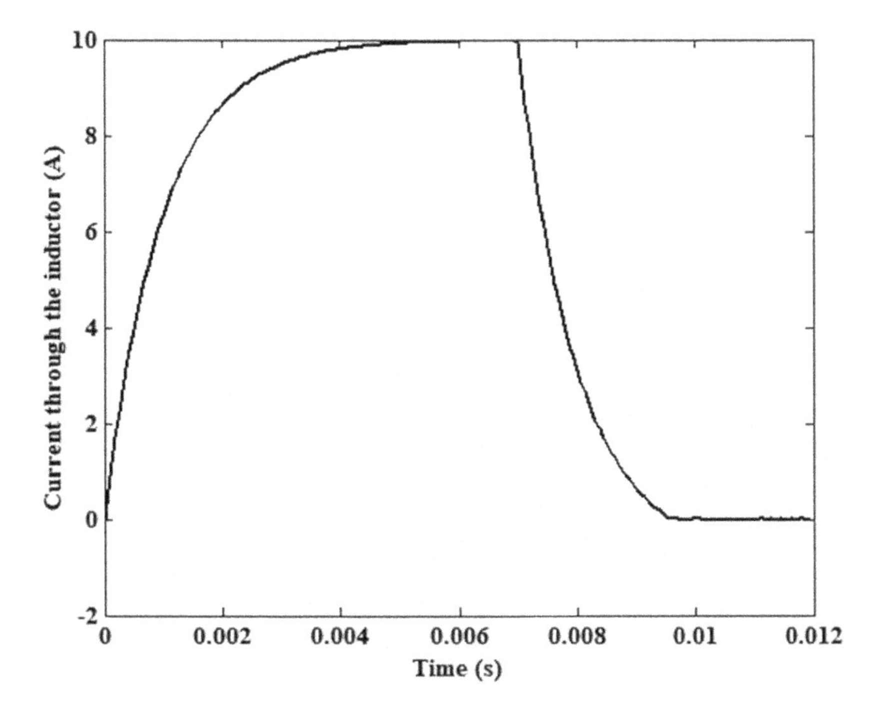

Figure 2.18 Plot of current through the inductor (A).

t (s) graph gives the energy associated with the inductor; and p_l (W) is given by Eqn. (2.15),

$$p_l = v_l \times i_l$$

where v_l—instantaneous value of voltage across the inductor (V); i_l—instantaneous value of current through the inductor (A).

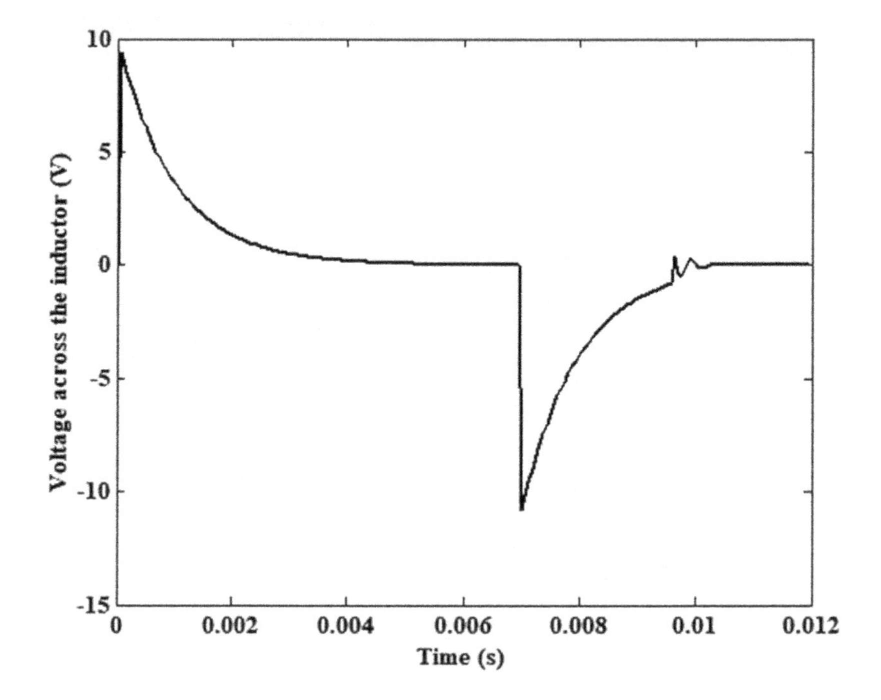

Figure 2.19 Plot of voltage across the inductor (V).

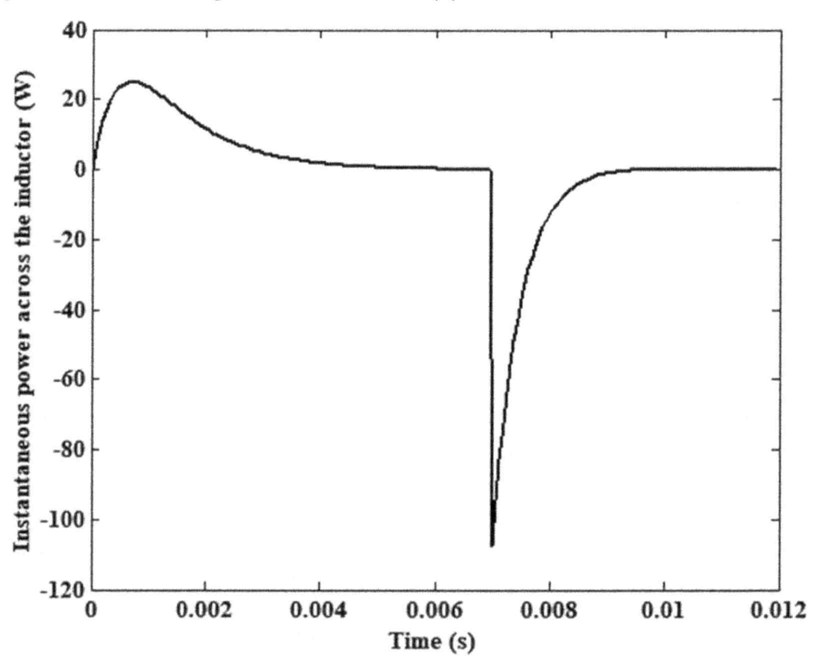

Figure 2.20 Plot of instantaneous power (W).

2.3.1 State-space approach

One may employ the state-space approach to model the dynamic behaviour of a circuit; the behaviour of a circuit can be fully defined by an input variable and a state variable as follows using general state-space representation:

$$\frac{dx}{dt} = A_s x + B_s u$$

$$y = C_s x + D_s u$$

where u—the input variable; x—the state variable; y—the output variable.

In the state-space approach, variables associated with energy sources are treated as input variables, and the variables associated with energy storing elements are treated as state variables. Energy sources can be voltage or current sources whereas energy storing elements can be inductance or capacitance.

One may note that kinetic energy (KE) stored in an inductor is given by

$$KE = \frac{1}{2} L \times i_l^2$$

where L—inductance (H); i_l—current through the inductor (A).

and potential energy (PE) stored in a capacitor is given by

$$PE = \frac{1}{2} C \times v_c^2$$

where C—capacitance (F); v_c—potential difference across the capacitor (V).

One may note that kinetic energy is stored in an inductor by the virtue of flow of current passing through the inductor, and potential energy is stored in a capacitor by the virtue of voltage across the capacitor. Thus, one may consider the current i_l as the state variable for an inductor and the voltage v_c as the state variable for a capacitor.

2.3.2 R-L circuit topology for state-space approach

Consider the R-L circuit shown in Figure 2.21, one may note that there is only one energy storage element which is nothing but the inductor and the current through the inductor i_l is the associated state variable. As shown in Figure 2.21, the variables associated with the input (V_s), state variable (i_l) and outputs under consideration (i_l and v_l) are only indicated in the topology for the state-space approach.

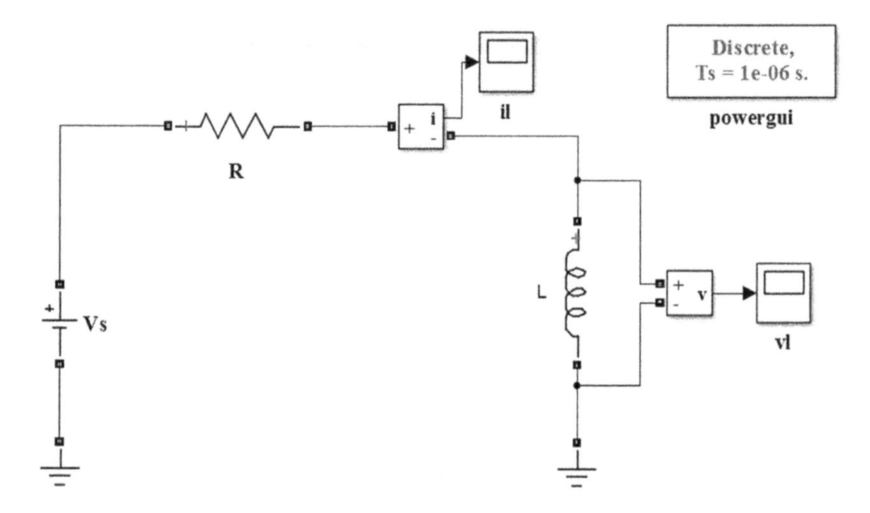

Figure 2.21 R-L circuit topology for state-space approach.

2.3.3 Mathematical formulation to obtain state equation

The voltage across the inductor in Figure 2.21 could be expressed as:

$$L\frac{di_l}{dt} = V_s - i_l R$$

where R—resistance (Ω); L—inductance (H).

rewriting Eqn. (2.20) in the general matrix state equation form one may get,

$$\left[\frac{di_l}{dt}\right] = [-R/L][i_l] + [1/L][V_s]$$

comparing Eqn. (2.16) and Eqn. (2.21) one may get,

$$[A_s] = [-R/L]$$

and

$$[B_s] = [1/L]$$

Taking i_l and v_l as the outputs, the respective output equations are

$$[i_l] = [1][i_l] + [0][V_s]$$

$$[v_l] = [-R][i_l] + [1][V_s]$$

In matrix form one may combine Eqn. (2.24) and Eqn. (2.25) to form a multi-output system as follows:

$$\begin{bmatrix} i_1 \\ V_1 \end{bmatrix} = \begin{bmatrix} 1 \\ -R \end{bmatrix}[i_1] + \begin{bmatrix} 0 \\ 1 \end{bmatrix}[V_s]$$

Comparing Eqn. (2.17) and Eqn. (2.26) one may get,

$$C_s = \begin{bmatrix} 1 \\ -R \end{bmatrix}$$

$$D_s = \begin{bmatrix} 0 \\ 1 \end{bmatrix}$$

and

Thus, the dynamic equation Eqn. (2.21) and output equation Eqn. (2.26) of the R-L circuit should be formulated to form the state equation as follows:

$$\begin{bmatrix} \frac{di_1}{dt} \end{bmatrix} = \begin{bmatrix} -R/L \end{bmatrix}[i_1] + \begin{bmatrix} 1/L \end{bmatrix}[V_s]$$

$$\begin{bmatrix} i_1 \\ V_1 \end{bmatrix} = \begin{bmatrix} 1 \\ -R \end{bmatrix}[i_1] + \begin{bmatrix} 0 \\ 1 \end{bmatrix}[V_s]$$

Eqn. (2.29a) implies that state vector

$$[x] = [i_1]$$

parameter matrix

$$[A_s] = [-R/L]$$

input vector

$$[u] = [V_s]$$

and input matrix

$$[B_s] = [1/L]$$

whereas Eqn. (2.29b) implies that output vector

$$[y] = \begin{bmatrix} i_1 \\ V_1 \end{bmatrix}$$

output matrix

$$[C_s] = \begin{bmatrix} 1 \\ -R \end{bmatrix}$$

feedforward matrix

$$[D_s] = \begin{bmatrix} 0 \\ 1 \end{bmatrix}$$

2.3.4 MATLAB codes to plot the charging characteristics of an R-L circuit using state-space approach

The MATLAB code given in mcode2–1 is used to generate the charging characteristics of an R-L circuit using state-space approach with the MATLAB step command. One may save mcode2–1 as an. m file, debug for any errors and run the program in MATLAB. Three different sets of values of resistance and inductance are chosen to compare the step response.

mcode2–1

```
>> R = 1; L = 0.5e-3;
>> As = [-R/L]; Bs = [1/L]; Cs = [1;-R]; Ds = [0;1];
>>step(As, Bs, Cs, Ds)
>> hold on
>> R = 1; L = 1e-3;
>> As = [-R/L]; Bs = [1/L]; Cs = [1;-R]; Ds = [0;1];
>>step(As, Bs, Cs, Ds)
>> R = 1; L = 2e-3;
>> As = [-R/L]; Bs = [1/L]; Cs = [1;-R]; Ds = [0;1];
>>step(As, Bs, Cs, Ds)
```

2.3.5 Output responses using state-space approach

The MATLAB code given in mcode2–1 generated the nature of inductor current and the voltage across the inductor when a unit step input is applied to the circuit. One may note that the time constant (L/R), 0.0005 s associated with the first set of values of resistance and inductance (para#1) is less than that of 0.001 s the time constant associated with para#2 and 0.002 s the time constant associated with para#3; as shown in Figure 2.22, the inductor current in the first case reaches final steady state faster than that of second and third. Analysis of Figure 2.22 shows that as the inductor current increases exponentially the voltage across the inductor decreases in the same fashion due to the voltage drop across the resistor.

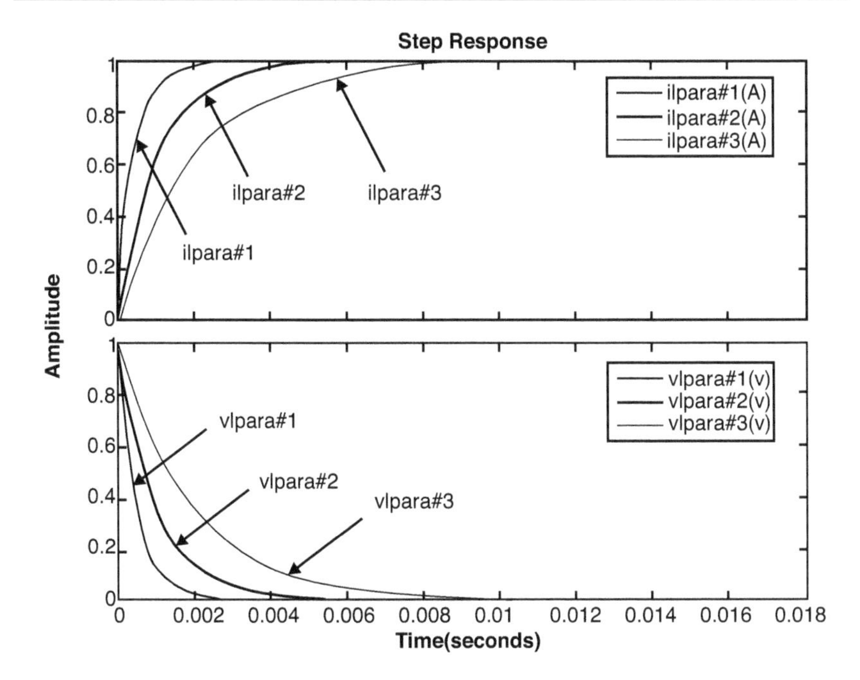

Figure 2.22 Output responses of an R-L circuit using state-space approach.

BIBLIOGRAPHY

Akagi H., Watanbe E. H., and Aredes M., Instantaneous power theory and applications to power conditioning (1st ed.). USA: John Wiley & Sons Inc., 2007.

Bradley D. A., Power electronics (2nd ed.). UK: Chapman & Hall, 1995.

George M., Advanced active power line conditioners for power quality improvement. Malaysia: Multimedia University, 2009.

George M., Basu K. P., and Pasupuleti J., Analysis of three-phase instantaneous power using MATLAB/PSB (1st ed.). Germany: LAP LAMBERT Academic Publishing, 2012.

George M. and Seen C. L., Modeling and control of zero-sequence current of parallel three-phase converters using MATLAB/PSB. USA: Power Systems Conference and Exposition IEEE PES October 11, 2004.

Hadi S., Power system analysis (2nd ed.). Singapore: McGraw-Hill Education (Asia), 2004.

Manke B. S., Linear control systems with MATLAB applications (8th ed.). India: Khanna Publishers, 2005.

Nagoor K. A., Control systems (1st ed.). India: R B A Publications, 1998.

Part-Enander E., Sjoberg A., Melin B., and Ishaksson P., The MATLAB handbook (1st ed.). USA: Addison Wesley, 1996.

The MathWorks Inc., Power system blockset for use with Simulink. USA: The MathWorks Inc., 2000.

Theraja A. K. and Theraja B. L., A textbook of electrical technology (25th ed.). India: S. Chand & Company Ltd., 2008.
Umanand L., Switched mode power conversion, online http://nptel.ac.in/, 2014.
Venkatesh B., Lecture notes on introduction to power system analysis. Malaysia: Multimedia University, 2004.
Wildi T., Electrical machines, drives, and power systems (4th ed.). USA: Prentice Hall International Inc., 2000.

Chapter 3

Analysis of system behaviour using transfer function model

3.1 INTRODUCTION

Stability analysis provides an insight into the system behaviour. For any system to be stable, oscillations in the output response should die out as fast as possible to reach the steady state final value. A transfer function model using the Laplace transform of the input and output in the s-plane lays the foundation for quickly predicting and graphing system behaviour from the stability point of view.

3.2 LAPLACE DOMAIN REPRESENTATION

Laplace transform transforms the time varying differential equations describing the dynamics of the system into algebraic equations in the s-domain/Laplace domain. The variable s represents exponentially growing/decaying sinusoidal waves; the growing/decaying rate is calculated by the modulus (σ) and the frequency of sinusoidal oscillations is represented by (ω). In the s-domain/Laplace domain the modulus (σ) is taken along the X-axis and the frequency (ω) is taken along the Y-axis; and the variable s is expressed as $s = \sigma + j\omega$. Transfer function of a system is the system gain G(s) in the Laplace domain as shown in Figure 3.1 and is defined as the Laplace transform of the output variable to the Laplace transform of the input variable ($G(s) = C(s)/R(s)$); one may note that the transfer function approach assumes initial conditions as zero.

3.2.1 Representation of an integrator in the s-domain/laplace domain

An integrator circuit is used in many engineering applications to accumulate the input over a defined time interval; the integrator function in the time-domain shown in Figure 3.2 could be represented in the s-domain/Laplace domain as shown in Figure 3.3.

DOI: 10.1201/9781003511236-3

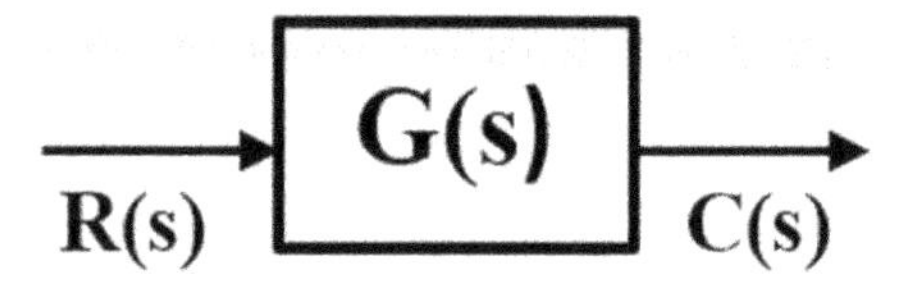

Figure 3.1 Representation of an input-output relation in the s-plane.

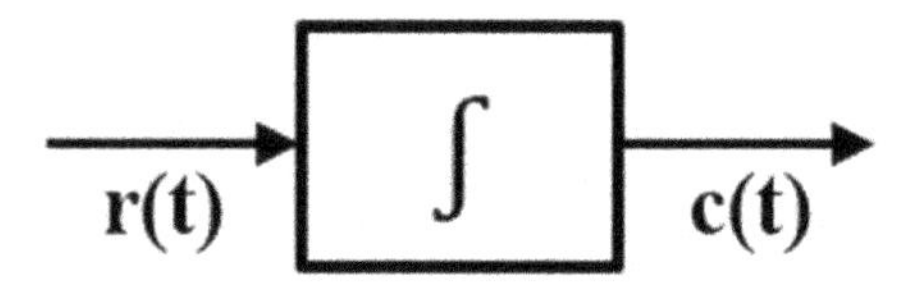

Figure 3.2 Time domain representation of an integrator.

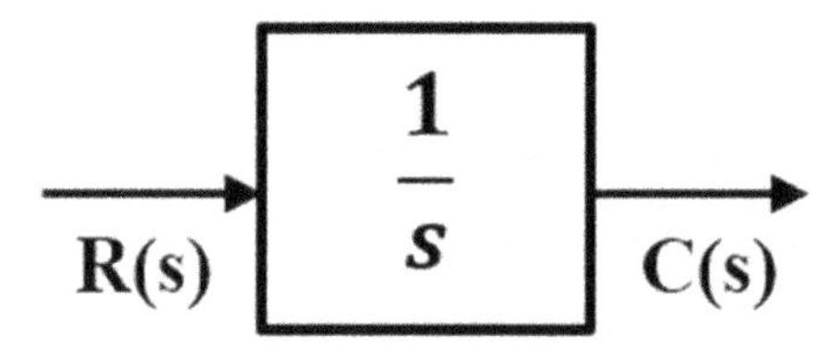

Figure 3.3 Laplace domain representation of an integrator.

3.2.2 Representation of a differentiator in the s-domain/laplace domain

A differentiator circuit is used in many engineering applications to produce the rate at which an input quantity varies with time. The differentiator function in the time-domain shown in Figure 3.4 could be represented in the s-domain/Laplace domain as shown in Figure 3.5.

3.2.3 Representation of a capacitor in the s-domain/laplace domain

A capacitive operation in the time domain shown in Figure 3.6 could be represented in the s-plane as shown in Figure 3.7.

Thus, the transfer function of a capacitive circuit is

$$G(s) = \frac{V_C(s)}{I_C(s)} = \frac{1}{sC} \tag{3.1}$$

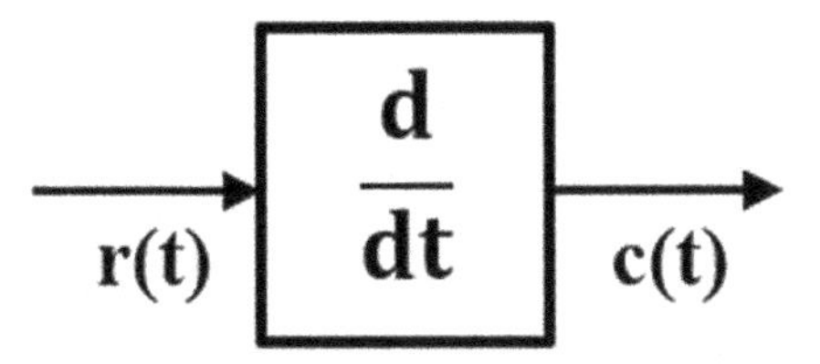

Figure 3.4 Time domain representation of a differentiator.

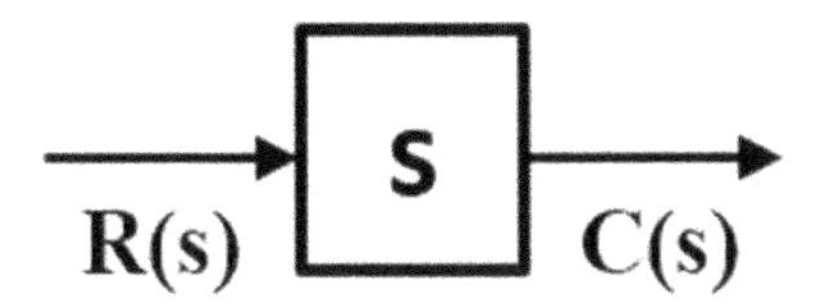

Figure 3.5 Laplace domain representation of a differentiator.

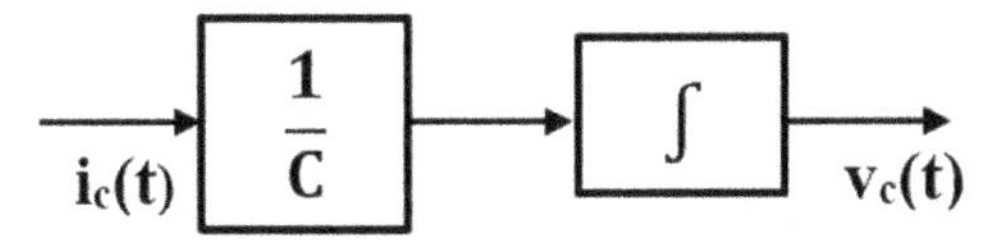

Figure 3.6 Time domain representation of a capacitive operation.

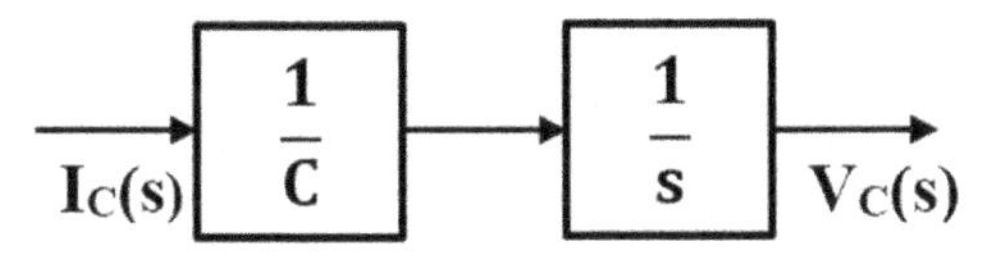

Figure 3.7 Laplace domain representation of a capacitive operation.

3.2.4 Representation of an inductor in the s-domain/ laplace domain

An inductive operation in the time-domain represented in Figure 3.8 could be represented in the s-plane as shown in Figure 3.9.

Thus, the transfer function of an inductive circuit is

$$G(s) = \frac{V_L(s)}{I_L(s)} = sL$$

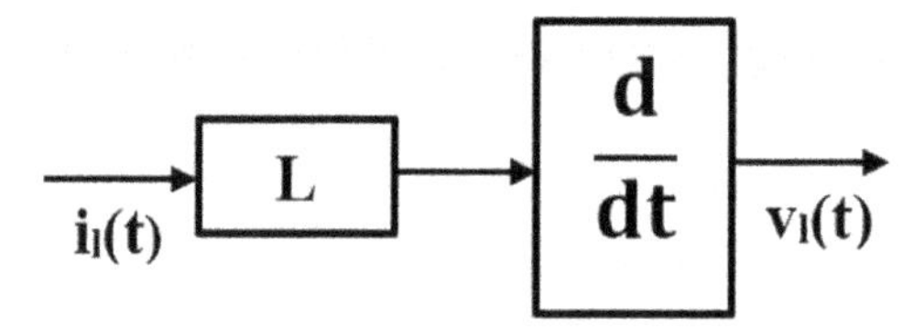

Figure 3.8 Time domain representation of an inductive operation.

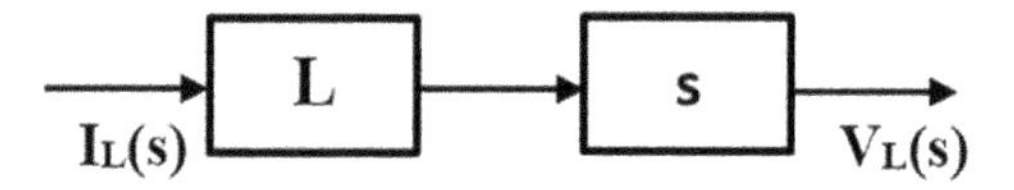

Figure 3.9 Laplace domain representation of an inductive operation.

3.2.5 Representation of an R-L-C circuit in the s-domain/laplace domain

One may draw the Laplace domain representation of an inductive-capacitive circuit feeding a load of resistance R as shown in Figure 3.10.

Applying Kirchoff's current law (KCL), the current through the inductor could be expressed as:

$$\frac{V_s(s) - V_c(s)}{sL} = \frac{V_c(s)}{1/sC} + \frac{V_c(s)}{R} \tag{3.3}$$

$$\frac{V_s(s)}{sL} = \frac{V_c(s)}{sL} + \frac{V_c(s)}{1/sC} + \frac{V_c(s)}{R} \tag{3.4}$$

$$\frac{V_s(s)}{sL} = \frac{V_c(s)}{sL} + sCV_c(s) + \frac{V_c(s)}{R} \tag{3.5}$$

$$\frac{V_s(s)}{sL} = V_c(s)\left[\frac{1+s^2LC+s(L/R)}{sL}\right] \tag{3.6}$$

Thus, gain of the system could be expressed as:

$$G(s) = \frac{V_c(s)}{V_s(s)} = \frac{1}{1 + s^2LC + \frac{sL}{R}} = \frac{1/LC}{s^2 + \frac{s}{RC} + \frac{1}{LC}} \tag{3.7}$$

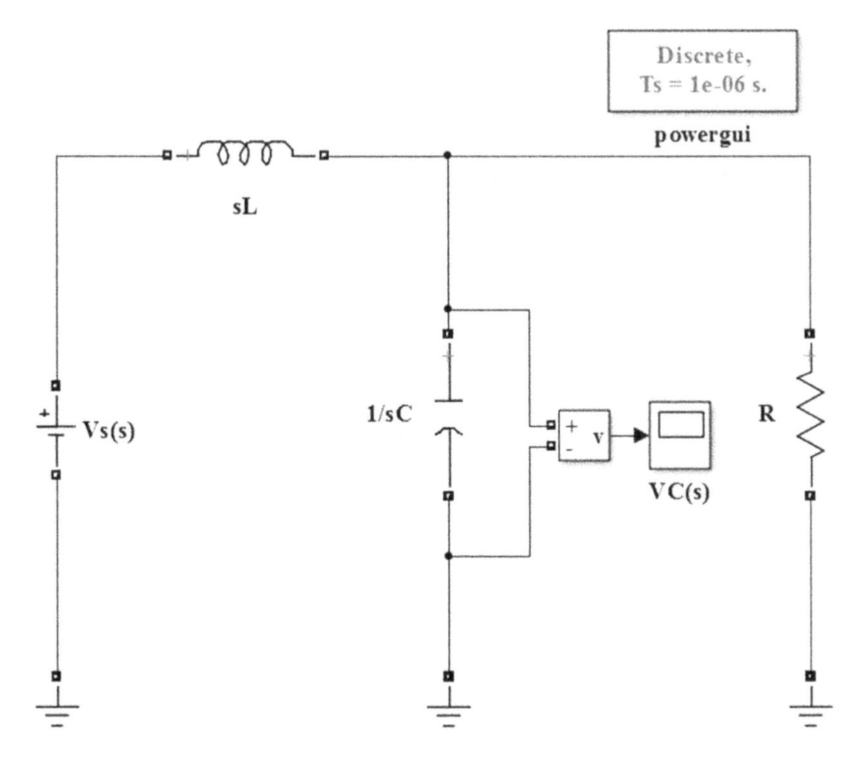

Figure 3.10 Laplace domain representation of an R-L-C circuit.

3.3 POLE-ZERO MAPPING IN THE S-DOMAIN/LAPLACE DOMAIN

Consider Figure 3.1, the transfer function (G(s) = C(s)/R(s)) explores an insight into the stability of the system by mapping the poles and zeros in the s-domain/Laplace domain.

3.3.1 Mapping of the poles

D(s), the roots of the denominator of the transfer function (G(s) = C(s)/R(s) = N(s)/D(s)) give the location of the poles; with poles p_1, p_2, ..., the denominator could be expressed as D(s) = ((s-p_1)(s-p_2) . . . . .). On the pole-zero map each pole is marked by the symbol (×); the gain of the system would be infinity at each pole or in other words the system is able to produce an output without any input signal.

3.3.2 Relationship between pole position and charging characteristics

Consider the Laplace domain representation of the R-C circuit shown in Figure 3.11, taking the capacitor voltage as the output, one may write,

$$G(s) = \frac{V_c(s)}{V_s(s)} = \frac{1/sC}{R + \frac{1}{sC}} = \frac{1}{sRC + 1} = \frac{1/RC}{s + 1/RC} \tag{3.8}$$

Analysis of Eqn. (3.8) shows that there is a pole at s = -1/RC; one may calculate the pole position for different values of parameters specified in Table 2.1 as follows:

para#1 R = 10 Ω; C = 10μF the pole is located at
$s = -1/RC = -1/(10 \times 10 \times 10^{-6}) = -10000$
para#2 R = 10 Ω; C = 100μF the pole is located at
$s = -1/RC = -1/(10 \times 100 \times 10^{-6}) = -1000$
para#3 R = 100 Ω; C = 100μF the pole is located at
$s = -1/RC = -1/(100 \times 100 \times 10^{-6}) = -100$

Comparison of the pole-zero map shown in Figure 3.12 with the charging characteristics illustrated in Figure 2.12 for different pole locations shows that if the pole position is away from the origin, the time constant is lesser and capacitor would charge faster as compared to poles closer to the origin.

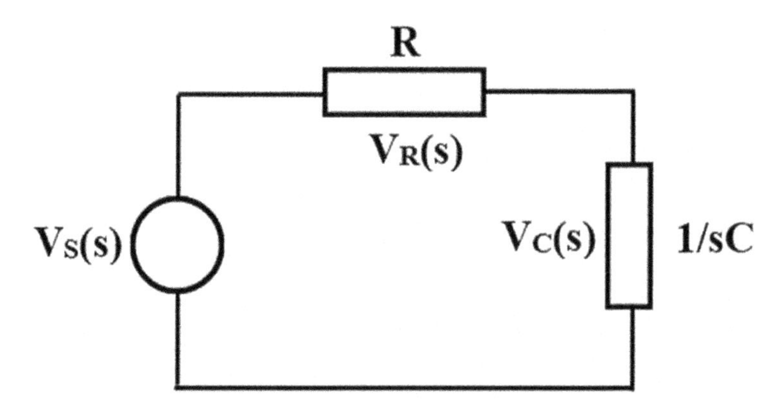

Figure 3.11 Laplace domain representation of an R-C circuit.

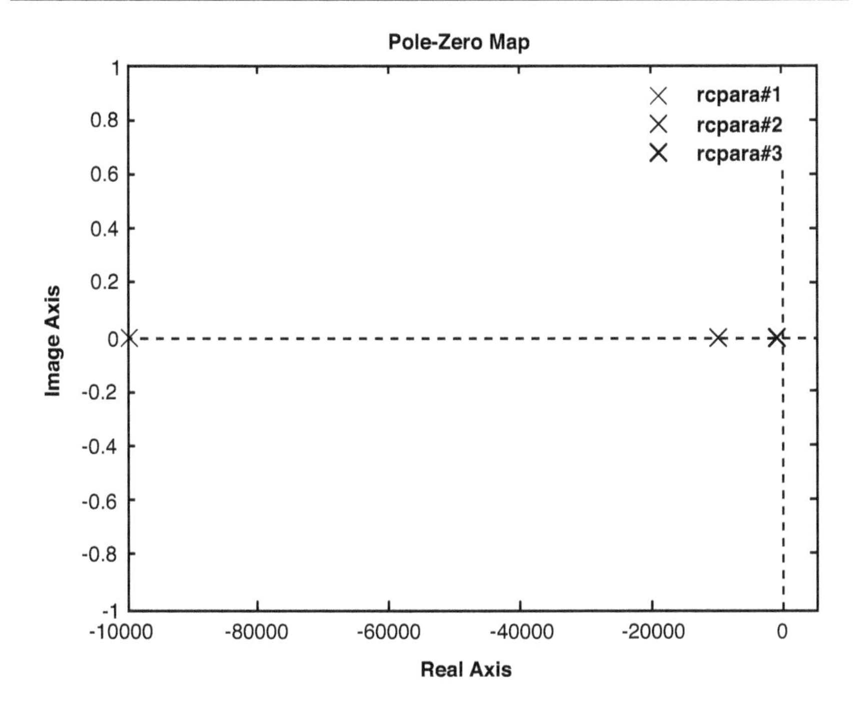

Figure 3.12 Pole-zero map of an R-C circuit for different parameter values.

3.3.3 Physical aspects of poles

One may note that the product RC is the time constant τ of an R-C circuit; meaning that the gain of the system would be infinity when

$$s = -\frac{1}{\tau} \tag{3.9}$$

or in other words

$$\sigma + j\omega = \frac{-1}{RC} \tag{3.10}$$

As the values of R and C are real numbers, the complex part $j\omega = 0$; meaning that the real part $\sigma = -1/RC$; or in otherwards the capacitor would reach a steady state value exponentially as given by Eqn. (2.13) without any oscillations.

When $s = -1/\tau$ the gain of the system is infinity meaning that even with $V_S(s) = 0$ there would be an output voltage across the capacitor. Meaning that even if the source voltage is shorted, the capacitor acts as a source and resistor acts as a load as illustrated in Figure 3.13.

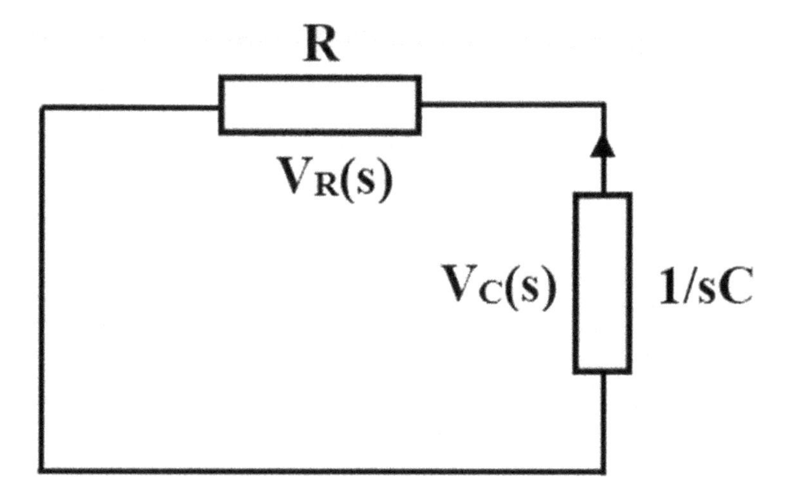

Figure 3.13 Illustration of an R-C circuit with infinite gain.

3.3.4 Mapping of the zeros

The roots of the numerator N(s) of the transfer function (G(s) = C(s)/R(s) = N(s)/D(s)) give the location of the zeros where the gain of the system is zero; thus, with zeros at z_1, z_2 . . . . . . , the numerator could be expressed as N(s) = $((s-z_1)(s-z_2)\ldots)$. On the pole-zero map each zero is marked by the symbol (o); the gain of the system would be zero at each zero or in other words the system is unresponsive for any magnitude of input.

3.3.5 Physical aspects of zeros

For example, consider the Laplace domain representation of an R-C high pass filter circuit shown in Figure 3.14, taking the voltage across the resistor as the output, one may write:

$$G(s) = \frac{V_{R(s)}}{V_{s(s)}} = \frac{R}{R + \frac{1}{sC}} = \frac{sRC}{sRC + 1} = \frac{s}{s + \frac{1}{RC}} \tag{3.11}$$

Analysis of Eqn. (3.11) shows that there is a zero at $s = 0$ and a pole at $s = -1/RC$. A zero at $s = 0$ or $\sigma + j\omega = 0 + j0$ means that an exposure to a signal which doesn't grow/decay without any oscillations and which in fact represents a constant DC signal. Thus, an R-C high pass filter circuit blocks any steady DC signal reaching its output.

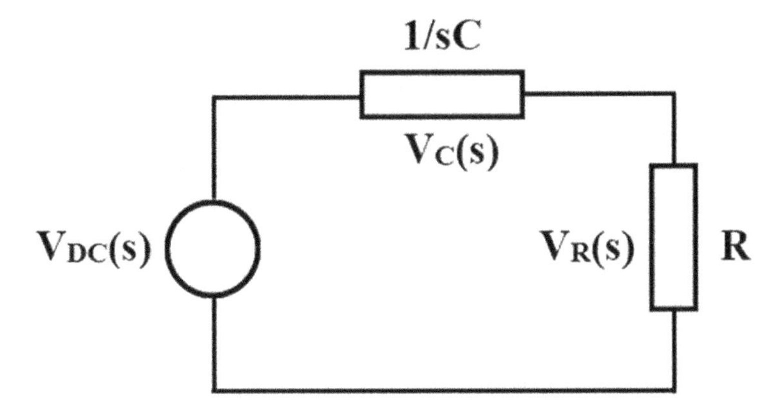

Figure 3.14 Illustration of an R-C circuit with zero gain.

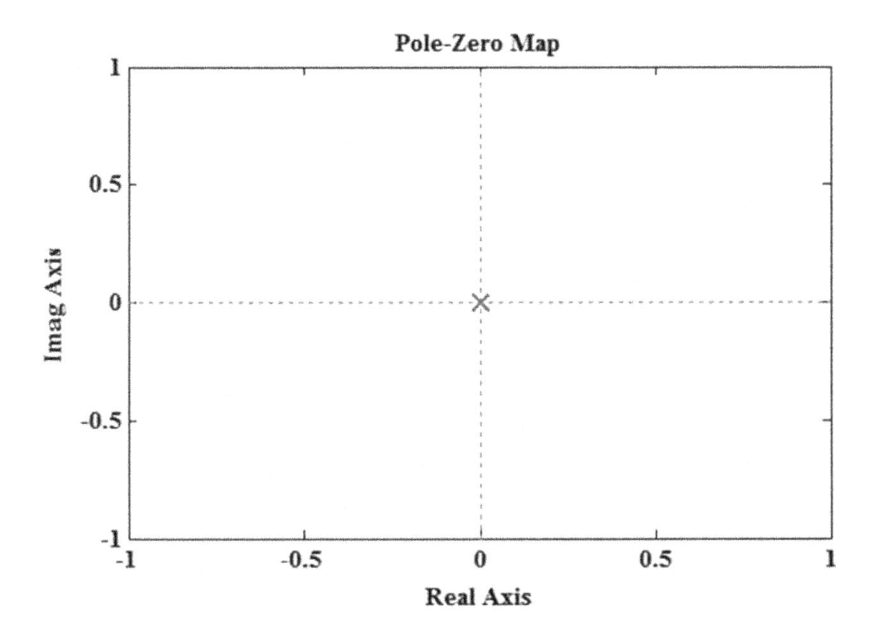

Figure 3.15 Pole-zero map of an integrator.

3.3.6 Pole-zero mapping using MATLAB® command

One may use the MATLAB command, pzmap(tf([1],[1 0])) to plot the pole-zero map shown in Figure 3.15 of an integrator $G(s) = 1/s = 1s^0/(1s^1 + 0s^0)$; note that there is no zero and there is only one pole at the origin $0 + j0$.

3.3.7 Pole-zero map of an R-L-C circuit

Consider the R-L-C circuit shown in Figure 3.10 with

$$L = 0.5\ \text{mH},\ C = 100\ \mu\text{F},\ R = 2\Omega$$

$$G(s) = \frac{1/LC}{s^2 + \frac{s}{RC} + \frac{1}{LC}} = \frac{1/\left(0.5\times10^{-3}\times100\times10^{-6}\right)}{s^2 + \frac{s}{2\times100\times10^{-6}} + \frac{1}{0.5\times10^{-3}\times100\times10^{-6}}}$$

$$G(s) = \frac{2\times10^7}{s^2 + 5000s + 2\times10^7} \tag{3.13}$$

$$= \frac{2\times10^7}{(s + (2500 - i3708))(s + (2500 + i3708))} \tag{3.14}$$

The pole-zero map of the R-L-C system represented by the transfer function model given by Eqn. 3.14 is shown in Figure 3.16; analysis of Figure 3.16 shows that there are two complex conjugate poles one at (–2500, j3708) and the other one at (–2500, –j3708) as given in Eqn. 3.14.

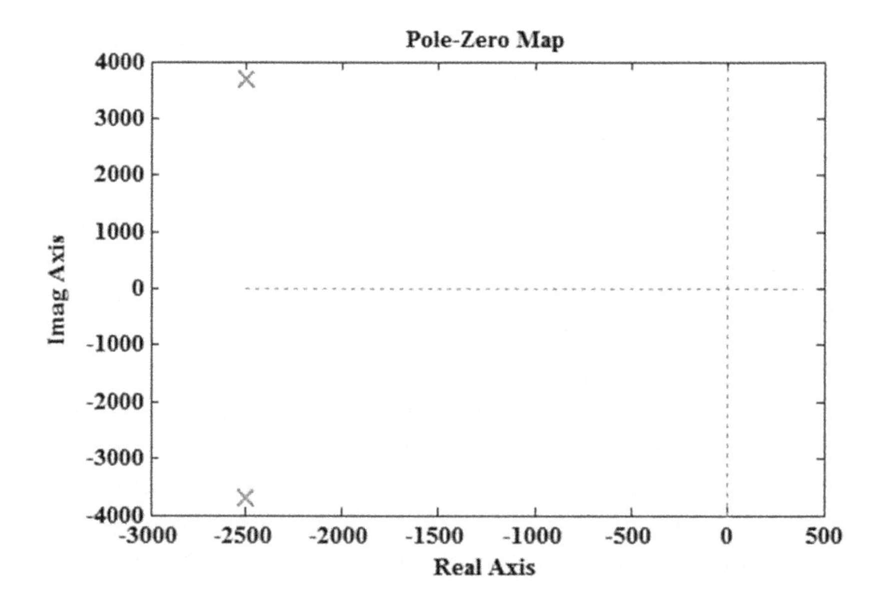

Figure 3.16 Pole-zero map of an R-L-C circuit.

3.4 SIGNIFICANT ASPECTS OF LOCATION OF POLES AND ZEROS FOR STABILITY ANALYSIS

Pole-zero map would be used to predict the stability of a system; for a system to be stable for any bounded input R(s), the output C(s) should be bounded; and then for a bounded input R(s), for the output C(s) to be bounded G(s) shouldn't have any poles on the right-hand side of the s-plane.

3.4.1 Stability analysis of a system with single pole at the origin

Consider an integrator system having a single pole at the origin $G(s) = C(s)/R(s) = 1/s$; meaning that the output response in the Laplace domain $C(s) = R(s)/s$; for a step input, $C(s) = (1/s) \times (1/s) = 1/s^2$, taking the Laplace inverse of C(s), $C(t) = L^{-1}(1/s^2) = t$; one may use the transfer function model as shown in Figure 3.17 to plot the step response of the integrator illustrated in Figure 3.18; one may note that the step input is changed from 0 to 1 at 0.02 s and the output response increases linearly with time.

For an integrator system as the pole is at the origin $s = 0 + j0$, any disturbance will decay at the rate of $e^{-(0)t}$; or in other wards the disturbance neither grows nor decays; the disturbance will just accumulate and if the input is zero the output will be held at the previous value as shown in Figure 3.19. Analysis of Figure 3.19 shows that the input of an integrator suddenly changed to zero at 0.05 s from an amplitude of one for a duration of 0.01 s and then returned to one; note that the output of the integrator remains at the previous value when the input is zero from 0.05 ms to 0.06 ms and the system is treated as marginally stable as long as the input won't contribute a double pole for the system output.

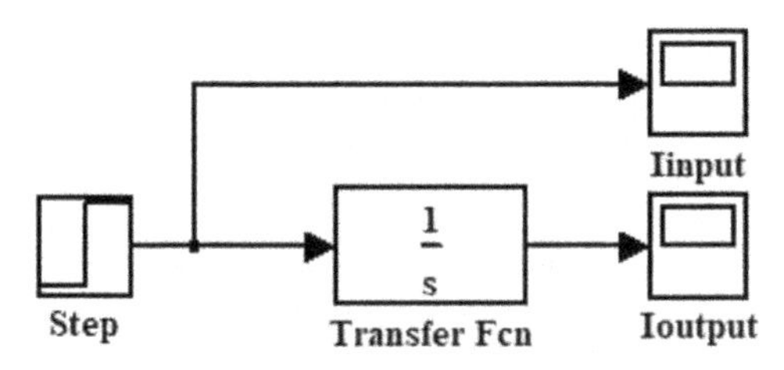

Figure 3.17 A transfer function model representing an integrator operation.

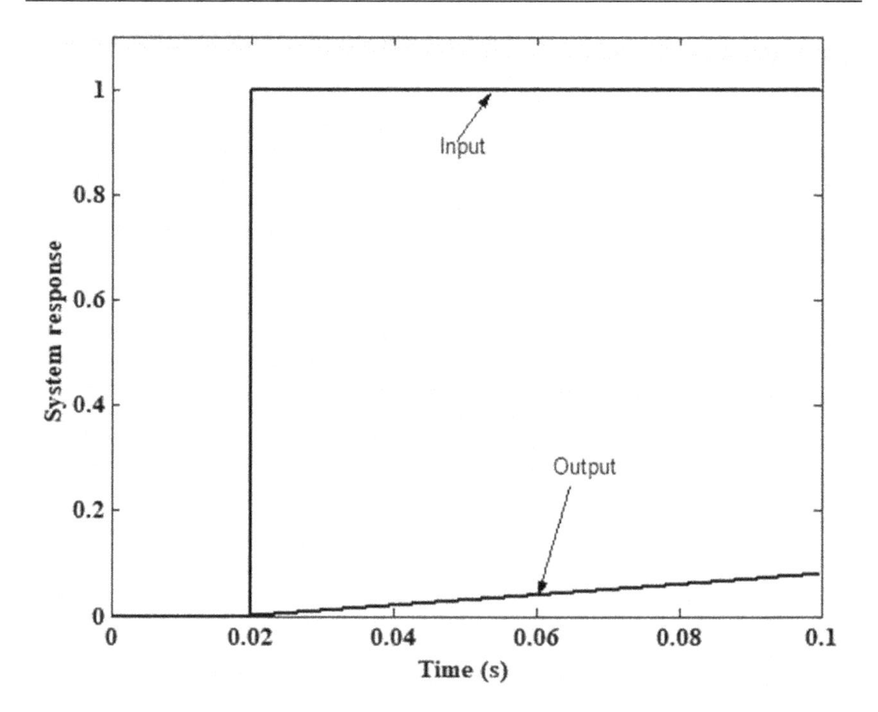

Figure 3.18 Step response of a system with single pole at the origin.

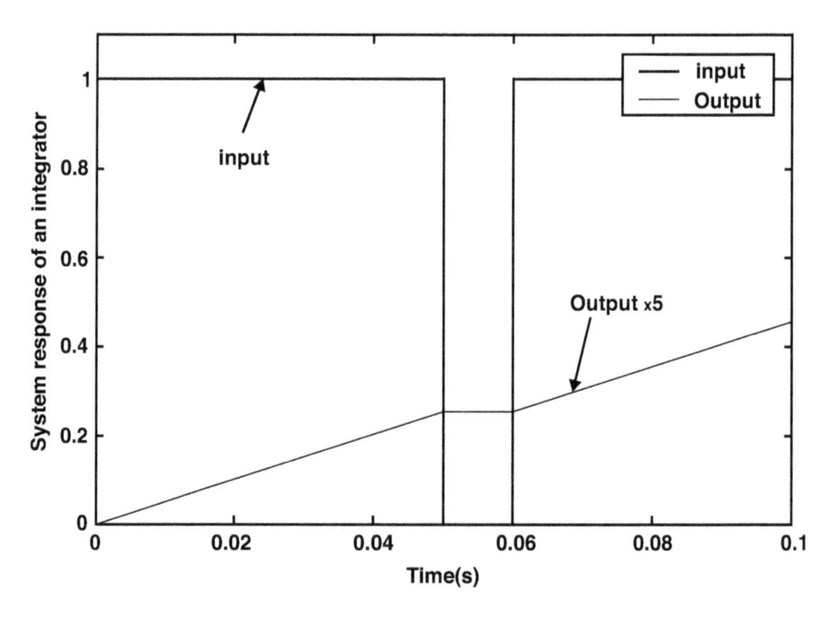

Figure 3.19 Stability analysis of an integrator.

3.4.2 Stability analysis of a system with complex conjugate poles on the imaginary axis

Consider a system $G(s) = \frac{C(s)}{R(s)} = \frac{s}{(s^2+1)} = \frac{s}{(s+i)(s-i)}$; meaning that there is a zero at the origin and there are complex conjugate poles on the imaginary axis one at +i and the other one at –i as shown in Figure 3.20. The output response of the system in the Laplace domain would be $C(s) = \frac{s}{(s^2+1)}R(s)$; for a step input, $C(s) = \frac{s}{(s^2+1)} \times \frac{1}{s}$, taking the Laplace inverse of C(s), $C(t) = L^{-1}\left(\frac{1}{s^2+1}\right) = \sin(t)$; step response of the system shown in Figure 3.21 shows that there is sustained sinusoidal oscillation in the output response if there are complex conjugate poles on the imaginary axis and the system is treated as marginally stable as long as the input won't contribute a double pole for the system output. If there are double poles on the imaginary axis the output would be growing sinusoidal oscillations.

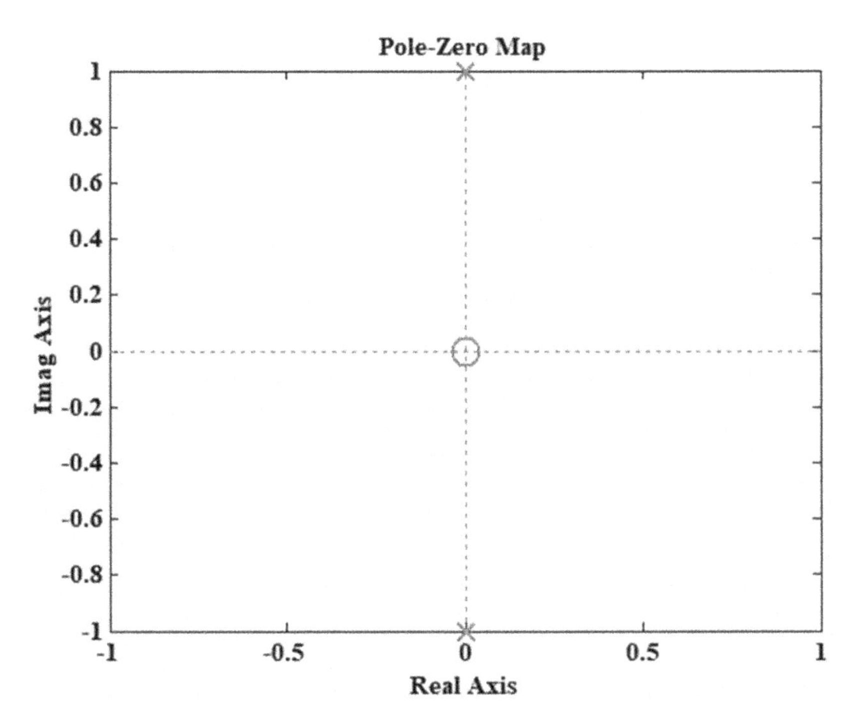

Figure 3.20 Pole-zero map of a system with poles on the imaginary axis.

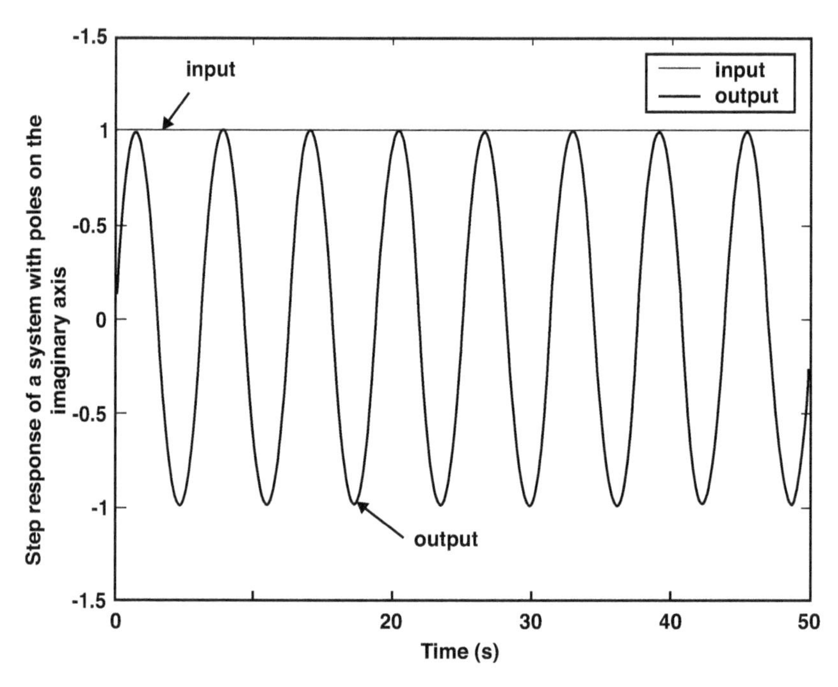

Figure 3.21 Step response of a system with complex conjugate poles on the imaginary axis.

3.4.3 Stability analysis of a system with complex conjugate poles on the left half of the s-plane

Consider a system $G(s) = \frac{C(s)}{R(s)} = \frac{1}{(s+1)^2 + 30^2}$; for the gain to be infinity $(s+1)^2 = 900i^2$; or in otherwards $s+1=\pm 30i$ or $s=-1\pm 30i$; meaning that there are complex conjugate poles on the left half of the s-plane, one at $-1+30i$ and the other one at $-1-30i$ as shown in Figure 3.22. Step response of the system shown in Figure 3.23 shows that the sinusoidal oscillations decay out in an exponential manner if the complex conjugate poles are on the left half of the s-plane and represents a stable system.

3.4.4 Stability analysis of a system with complex conjugate poles on the right half of the s-plane

Consider a system $G(s) = \frac{C(s)}{R(s)} = \frac{1}{(s-1)^2 + 30^2}$; pole-zero map shown in Figure 3.24 shows that there are complex conjugate poles on the right half of the s-plane one at $1+30i$ and the other one at $1-30i$. Step response of the system shown in Figure 3.25 shows that the sinusoidal oscillations grow in

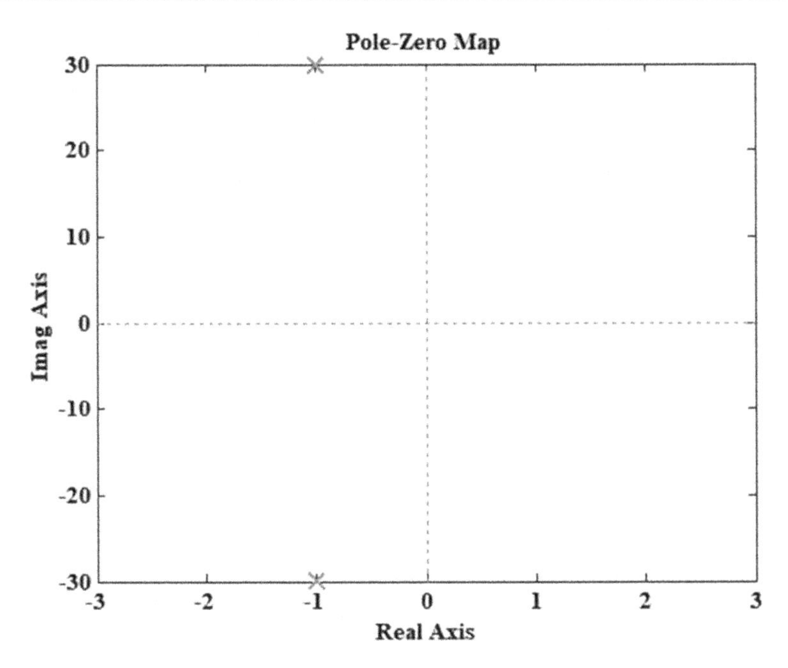

Figure 3.22 Pole-zero map of a system with complex conjugate poles on the left half of the s-plane.

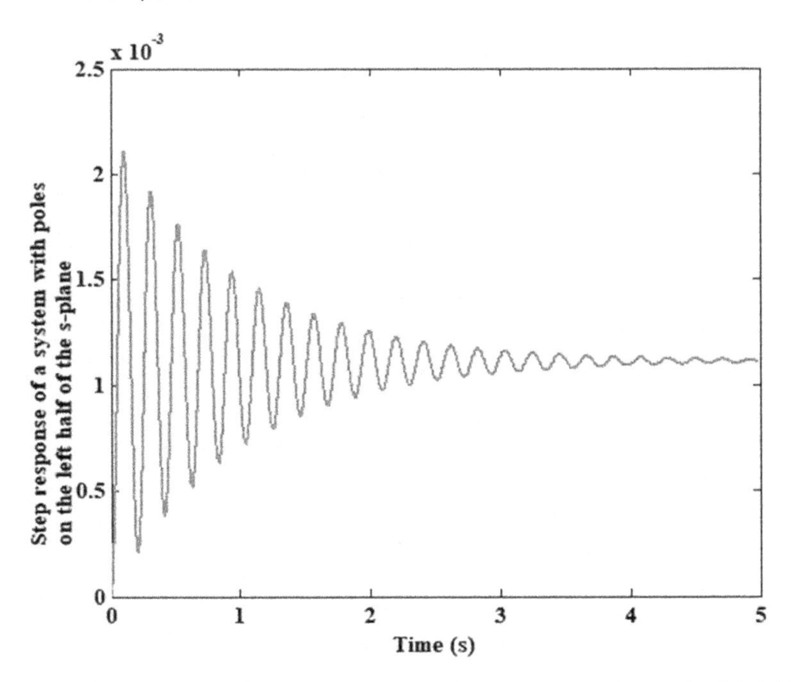

Figure 3.23 Step response of a system with complex conjugate poles on the left half of the s-plane.

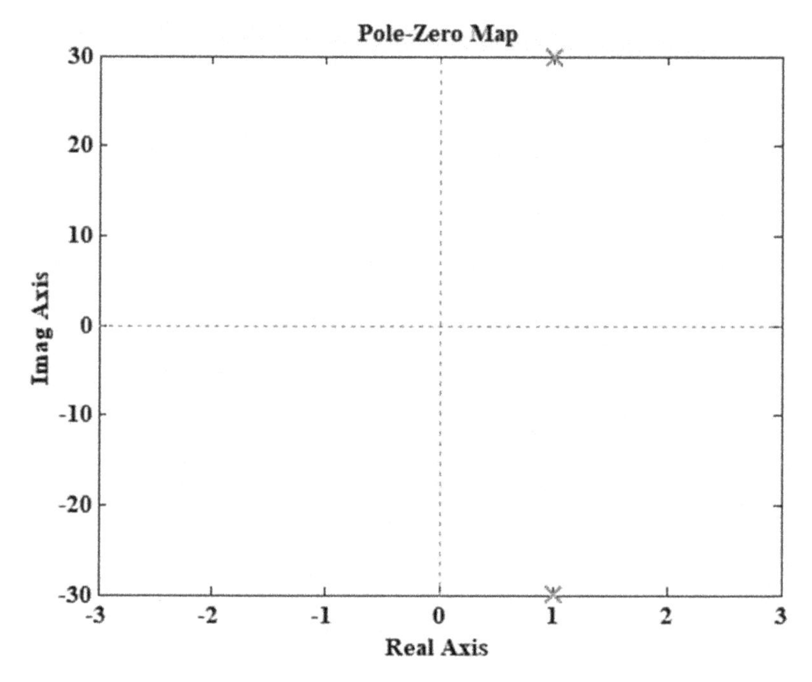

Figure 3.24 Pole-zero map of a system with complex conjugate poles on the right half of the s-plane.

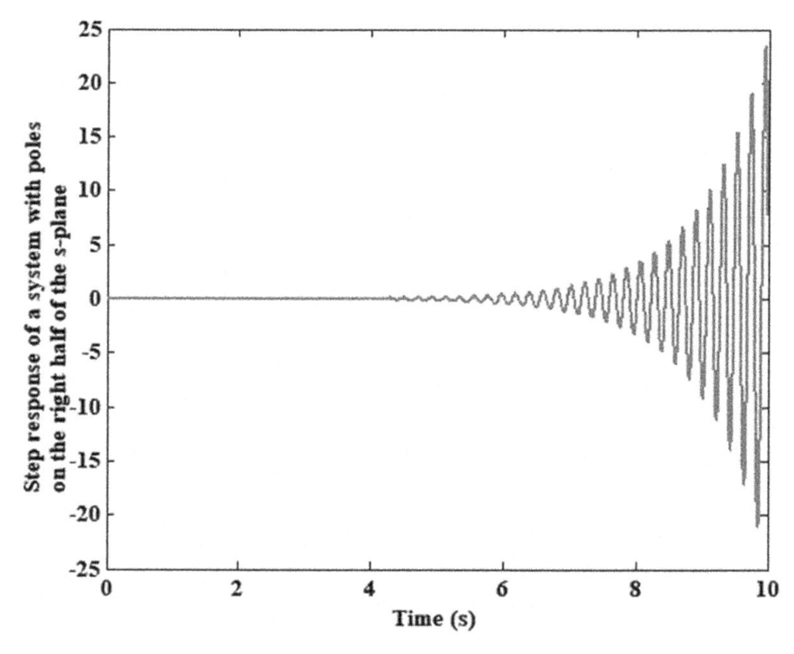

Figure 3.25 Step response of a system with complex conjugate poles on the right half of the s-plane.

an exponential manner if the complex conjugate poles are on the right half of the s-plane and represents an unstable system.

3.4.5 Stability analysis of a system with zeros and poles on the right half of the s-plane

Figure 3.26 shows the pole-zero map of a system having transfer function $G(s) = \frac{s^2 - 2s}{s^2 - 1}$. One may note that the system is having zeros at (0,0), (2,0); and poles at (1,0), (−1,0). As illustrated in Figure 3.27 the system is unstable.

3.5 MATHEMATICAL FORMULATION OF THE TIME RESPONSE OF AN R-L-C CIRCUIT

To analyze the time response using Laplace domain representation, one may consider the Eqn. 3.14 of Figure 3.10, one may note that Eqn. 3.14 could be written as:

$$V_C(s) = \frac{2 \times 10^7}{(s + (2500 - 3708i)(s + 2500 + 3708i))} V_S(s) \tag{3.15}$$

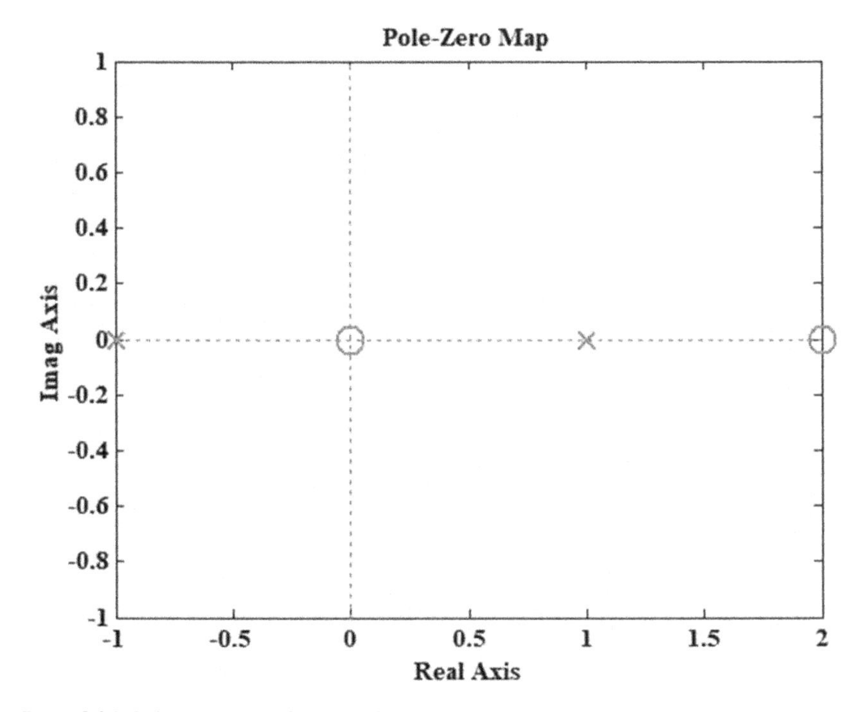

Figure 3.26 Pole-zero map of an unstable system.

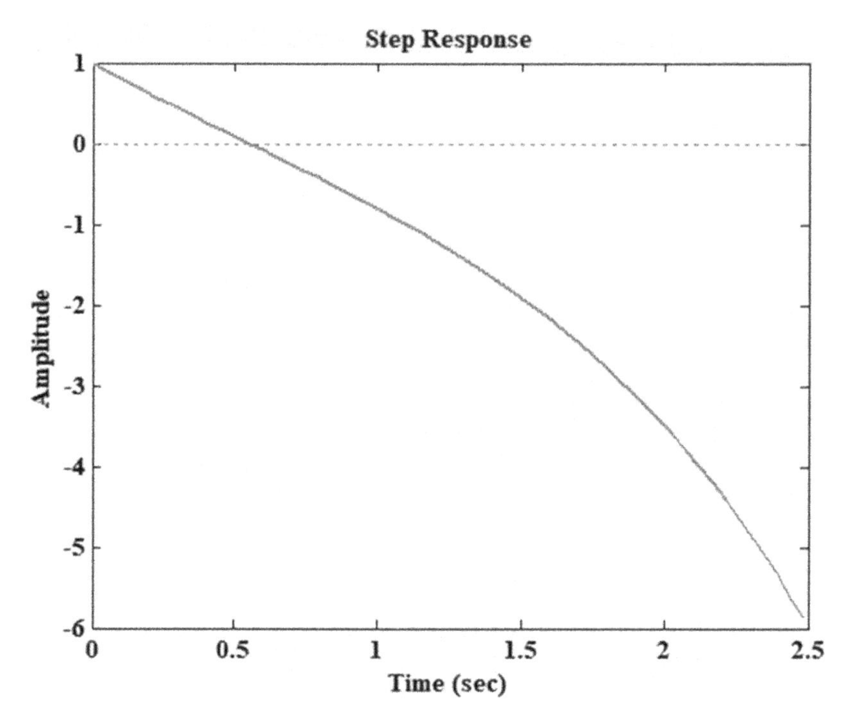

Figure 3.27 Step response of an unstable system.

With a step input,

$$V_C(s) = \frac{2\times10^7}{(s+(2500-3708i)(s+2500+3708i))}\times\frac{1}{s} \tag{3.16}$$

Applying inverse Laplace transform to express the voltage across the capacitor in the time domain.

$$\therefore v_c(t) = L^{-1}\left(\frac{2\times10^7}{s(s+(2500-3708i))(s+(2500+3708i))}\right) \tag{3.17}$$

Using partial fraction expansion, Eqn. 3.16 could be expressed as:

$$= \frac{A_1}{s}+\frac{A_2}{s+(2500-3708i)}+\frac{A_3}{(s+2500+3708i)} \tag{3.18}$$

Thus,

$$2\times10^7 = A_1\left(s^2+5000s+2\times10^7\right)+A_2s(s+2500+3708i)$$
$$+A_3s(s+2500-3708i)$$

Rearranging Eqn. 3.19, one may get,

$$2\times10^7 = 2\times10^7A_1 +$$
$$s\left(5000A_1+(2500+3708i)A_2+(2500-3708i)A_3\right)+s^2\left(A_1+A_2+A_3\right)$$

$$\therefore\left(2\times10^7\right)A_1 = 2\times10^7 \quad (3.21)$$

$$s\left(5000A_1+(2500+3708i)A_2+(2500-3708i)A_3\right)=0 \quad (3.22)$$

$$s^2\left(A_1+A_2+A_3\right)=0 \quad (3.23)$$

From Eqn. 3.21,

$$A_1 = 1 \quad (3.24)$$

From Eqn. 3.23,

$$\left(A_1+A_2+A_3\right)=0 \quad (3.25)$$

Substituting $A_1 = 1$ in Eqn. 3.25, one may get,

$$\left(1+A_2+A_3\right)=0 \quad (3.26)$$

$$A_3 = -\left(1+A_2\right) \quad (3.27)$$

Substituting Eqn. 3.24 and Eqn. 3.27 in Eqn. 3.22, one may get,

$$5000+2500A_2+3708iA_2-\left(2500-3708i+2500A_2-3708iA_2\right)=0 \quad (3.28)$$

Re-arranging Eqn. 3.28, one may get,

$$5000+2500A_2+3708iA_2-2500+3708i-2500A_2+3708iA_2=0 \quad (3.29)$$

From Eqn. 3.29, one may get,

$$5000+3708iA_2-2500+3708i+3708iA_2=0 \quad (3.30)$$

From Eqn. 3.30, one may get,

$$2500 + 2 \times 3708\text{i}A_2 + 3708\text{i} = 0$$

$$A_2 = -\frac{2500 + 3708\text{i}}{2 \times 3708\text{i}} \tag{3.32}$$

Substituting Eqn. 3.32 in Eqn. 3.27, one may get,

$$A_3 = -\left(1 - \frac{2500 + 3708\text{i}}{2 \times 3708\text{i}}\right) = -\left(\frac{2 \times 3708\text{i} - 2500 - 3708\text{i}}{2 \times 3708\text{i}}\right) \tag{3.33}$$

$$\therefore A_3 = -\left(\frac{3708\text{i} - 2500}{2 \times 3708\text{i}}\right) = \frac{2500 - 3708\text{i}}{2 \times 3708\text{i}} \tag{3.34}$$

Thus,

$$\therefore v_c(t) = L^{-1}\left(\left(\frac{1}{s}\right) - \left(\frac{2500 + 3708\text{i}}{2 \times 3708\text{i}(s + (2500 - 3708\text{i}))}\right) + \left(\frac{2500 - 3708\text{i}}{2 \times 3708\text{i}(s + (2500 + 3708\text{i}))}\right)\right)$$

Thus, the capacitor voltage in the time domain could be expressed as:

$$v_c(t) = 1 - \frac{2500 + 3708\text{i}}{2 \times 3708\text{i}} e^{-(2500 - 3708\text{i})t} + \frac{2500 - 3708\text{i}}{2 \times 3708\text{i}} e^{-(2500 + 3708\text{i})t}$$

$$v_c(t) = 1 + \left(-\frac{1}{2} + \frac{625}{1854} i\right) e^{(-2500 + 3708\text{i})t} + \left(-\frac{1}{2} - \frac{625}{1854} i\right) e^{(-2500 - 3708\text{i})t}$$

One may verify Eqn. 3.37 using MATLAB commands as follows:

```
>> format short G
  >> syms s t
  >> ilaplace((1/s),s,t)
  ans =
  1
  >> ilaplace(((2500+3708i)/((2*3708i)*(s + (2500 – 3708i)))),s,t)
  ans =
  (1/2 – 625/1854*i)*exp((–2500+3708*i)*t)
  >> ilaplace(((2500–3708i)/((2*3708i)*(s + (2500+3708i)))),s,t)
  ans =
  (–1/2 – 625/1854*i)*exp((–2500 – 3708*i)*t)
```

Thus one may get,

```
vc(t)=1–((1/2 – 625/1854*i)*exp((–2500 + 3708*i)*t))+(–1/2–
625/1854*i)*exp((–2500 – 3708*i)*t)
```

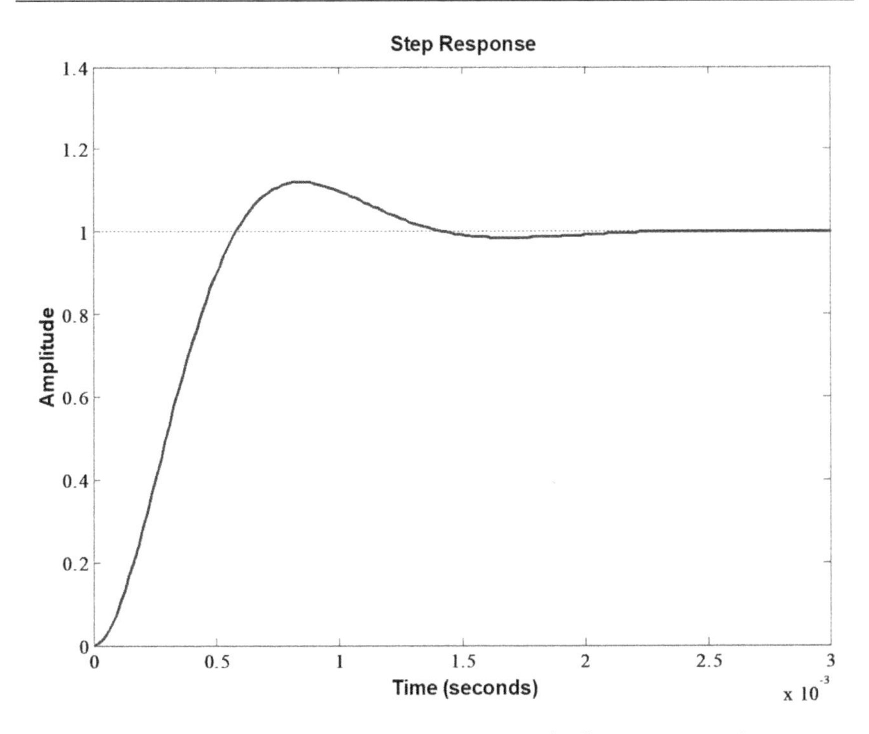

Figure 3.28 Step response of an R-L-C circuit using transfer function approach.

The step response of the R-L-C circuit shown in Figure 3.28 shows that the capacitor voltage reaches final steady state value of one at t = 1.5ms.

One may apply initial value theorem and final value theorem using transfer function representation to determine the initial and final values of the time varying variable $v_c(t)$.

From Eqn. 3.37, one may write

$$V_c(s) = \frac{2\times10^7}{s^2+5000s+2\times10^7}V_s(s) = \frac{2\times10^7}{\left(s^2+5000s+2\times10^7\right)}\times\frac{1}{s} \tag{3.38}$$

3.5.1 Initial value theorem

Applying initial value theorem, one may get:

$$\lim_{t\to0} f(t) = \lim_{s\to\infty}(sF(s)) = \lim_{s\to\infty}\left(s\times\frac{2\times10^7}{s\times\left(s^2+5000s+2\times10^7\right)}\right) \tag{3.39}$$

$$= \lim_{s\to\infty}\frac{2\times10^7}{\left(s^2+5000s+2\times10^7\right)} = 0 \tag{3.40}$$

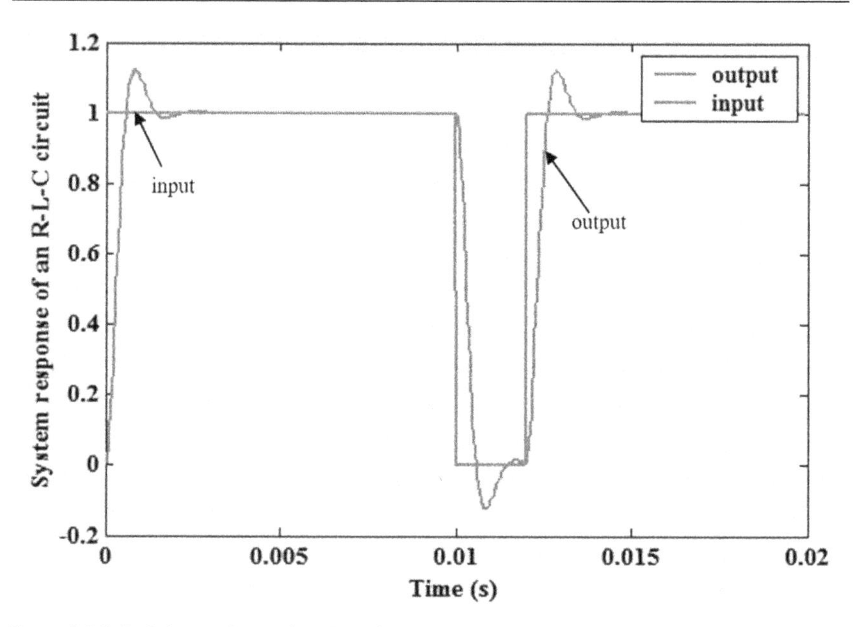

Figure 3.29 Stability analysis of an R-L-C system.

3.5.2 Final value theorem

Applying final value theorem, one may get

$$\lim_{t\to\infty} f(t) = \lim_{s\to 0}(sF(s)) = \lim_{s\to 0}\left(s \times \frac{2\times 10^7}{s \times \left(s^2 + 5000s + 2\times 10^7\right)}\right) = 1 \tag{3.41}$$

One may use the MATLAB code step(tf([2e7],[1 50000 2e7])) to plot the step response shown in Figure 3.28 using transfer function approach. Stability of the system is verified as shown in Figure 3.29 by introducing a disturbance for a short duration.

BIBLIOGRAPHY

Benjamin C. K., Automatic control systems (5th ed.). India: Prentice Hall of India Private Limited, 1989.

Dorf R. C. and Bishop R. H., Modern control systems (1st ed.). World student series. USA: Addison-Wesley, 1998.

Electrical systems/basic alternating (AC) theory/transfer function analysis, online http://control.com/textbook/ac-electricity/transfer-function-analysis/, 2022.

Hadi S., Power system analysis (2nd ed.). Singapore: McGraw-Hill Education (Asia), 2004.

Manke B. S., Linear control systems with MATLAB applications (8th ed.). India: Khanna Publishers, 2005.
Nagoor K. A., Control systems (1st ed.). India: R B A Publications, 1998.
Ogata K., Modern control engineering (3rd ed., International). USA: Prentice Hall International Inc., 1997.
Part-Enander E., Sjoberg A., Melin B., and Ishaksson P., The MATLAB handbook (1st ed.). USA: Addison Wesley, 1996.
Umanand L., Switched mode power conversion, online http://nptel.ac.in/, 2014.

Chapter 4

Stability analysis of closed loop systems

4.1 INTRODUCTION

The output performance of a plant varies with parameter variations; in an open loop system there is no control action to achieve the desired output performance with parameter variations; as shown in Figure 4.1, in an open loop system the output performance of a plant depends only on the system input; whereas in a closed loop control system, the actual output is compared with the desired output and the error signal is fed to a controller to achieve the desired performance as shown in Figure 4.2. However, there is a possibility of instability in the closed loop response even when the open loop system is stable. One may use a Bode plot for the stability analysis of closed loop systems.

4.2 BODE PLOT

Bode plots use frequency domain analysis; the response of a system to a sinusoidal input of constant magnitude (σ) but over a wide range of frequency (ω) is investigated in frequency domain analysis. The magnitude and phase of the gain with respect to frequency are plotted in the frequency domain analysis; the resulting plot is called a Bode plot. To illustrate the effect of wide range of frequency on the magnitude and phase of the gain, logarithmic scale in dB has been used.

4.2.1 Bode plot of an R-L circuit

One may consider the R-L circuit shown in Figure 2.21, the resulting current i_l through the circuit is given by

$$i_l = V_s\left(\frac{1}{R + j\omega L}\right) \tag{4.1}$$

DOI: 10.1201/9781003511236-4

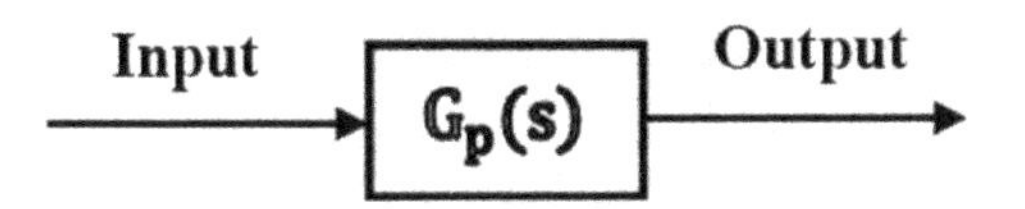

Figure 4.1 An open loop plant operation.

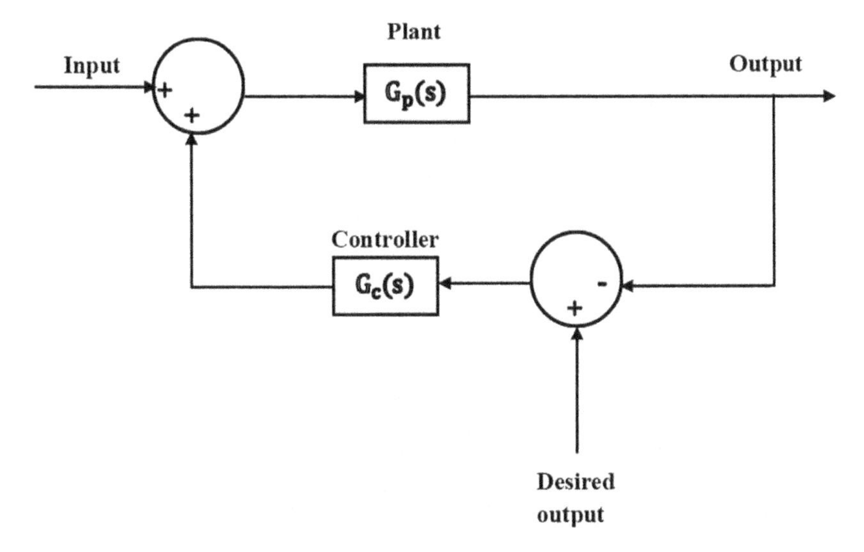

Figure 4.2 A closed loop plant operation.

The gain

$$G = \frac{i_1}{Vs} = \left(\frac{1}{R + j\omega L}\right) = \frac{1/R}{1 + j\omega L/R} \tag{4.2}$$

4.2.1.1 *Magnitude in dB*

From Eqn. 4.2, one may get magnitude of gain

$$|G| = \frac{1/R}{\sqrt{\left(1 + (\omega L/R)^2\right)}} = \frac{1/R}{\sqrt{\left(1 + (\omega\tau)^2\right)}} \tag{4.3}$$

where $\tau = L/R$ is the time constant of the circuit.

- Since the value of R is normally low, on the dB scale, the resistance has negligible effect on the gain.

- For $\omega < 1/\tau$, the term $(\omega L/R)^2$ is very small and can be neglected.
- For $\omega > 1/\tau$, term 1 in the denominator can be neglected.

To get a clear view of gain, every decade of frequency over a wider range is considered; one may calculate the gain in dB as follows:

$$\text{Magnitude of Gain} = 2\log|G|\,\text{bel} = 20\log|G|\,\text{dB} \tag{4.4}$$

Let,

$$R = 1\Omega; L = 1\text{mH; then } \frac{L}{R} = \tau = 0.001\text{s} \tag{4.5}$$

For values of

$$\omega \le \frac{1}{\tau} = \frac{1}{0.001} = 10^3 \Rightarrow |G(j\omega)| \Rightarrow 1;\ 20\log(1) = 0\text{dB} \tag{4.6}$$

For values of

$$\omega \ge 10^3 \Rightarrow |G(j\omega)| \Rightarrow \frac{1}{\omega\frac{L}{R}} \tag{4.7}$$

When the frequency is increased tenfold, say $\omega_2 = 10\omega_1$

$$20\log\left(\frac{1}{\omega_2}\right) = 20\log\left(\frac{1}{10\omega_1}\right) = 20\log(0.1) + 20\log\left(\omega_1^{-1}\right) = -20 + 20\log\left(\omega_1^{-1}\right) \tag{4.8}$$

Eqn. 4.8 implies that for every decade of increase in frequency, there is a 20dB decrease in gain.

4.2.1.2 Phase angle

From Eqn. 4.2, one may get phase angle of gain

$$\angle G = -\tan^{-1}\left(\frac{\omega L}{R}\right) \tag{4.9}$$

4.2.2 Magnitudes and phase angles of gains for different parameter values

Bode plots of an R-L circuit for three different parameter values are plotted as shown in Figure 4.3 using MATLAB® codes given in mcode4–1. From Eqn. 2.29a, one may get, $[A_s] = [-R/L]$ and $[B_s] = [1/L]$. Taking current i_l as the only

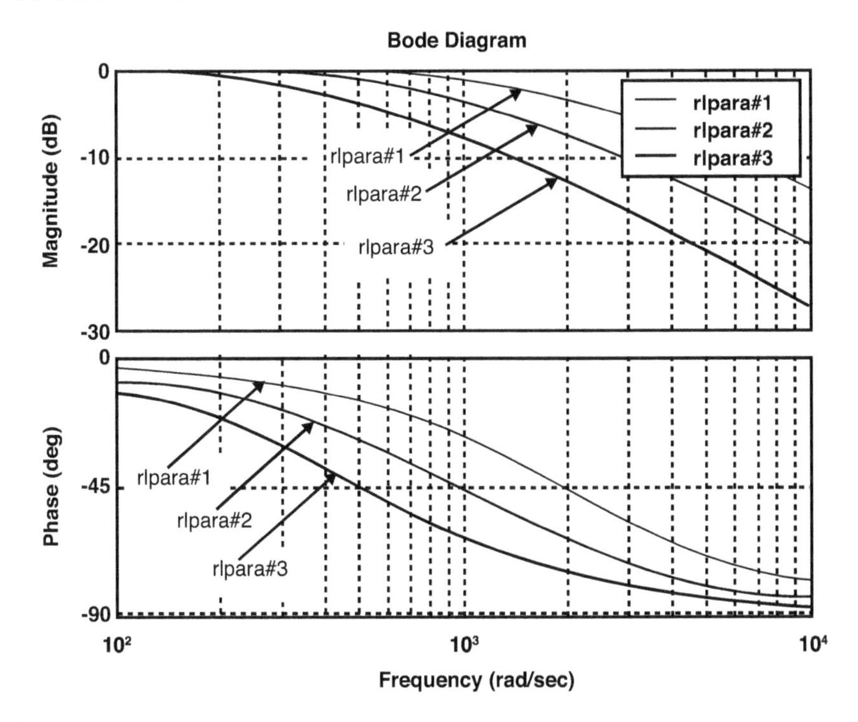

Figure 4.3 Bode plots of an R-L circuit for different parameter values.

one output, from Eqn. 2.29b, one may write, $[C_s] = [1]$ and $[D_s] = [0]$. One may determine the magnitudes of gains and the phase angles for different parameter values as seen in Table 4.1 and Table 4.2 respectively; one may note that frequency ω is varied from 1 rad/s to 10^6 rad/s.

mcode4–1

```
≫R = 1; L = 0.5e–3;
≫As = [–R/L]; Bs = [1 – L]; Cs = [1]; Ds = [0];
≫bode(As,Bs,Cs,Ds)
≫hold on
≫R = 1; L = 1e–3;
≫As = [–R/L]; Bs = [1 – L]; Cs = [1]; Ds = [0];
≫bode(As,Bs,Cs,Ds)
≫hold on
≫R = 1; L = 2e–3;
≫As = [–R/L]; Bs = [1 – L]; Cs = [1]; Ds = [0];
≫bode(As,Bs,Cs,Ds)
```

Table 4.1 Magnitudes of Gains in dB of an R-L Circuit for Different Parameter Values

para#1 R = 1Ω; L = 0.5mH; L/R = 0.5 ms			
ω (rad/s)	$\log_{10}(\omega)$	|G(jω)|	20 logG(jω)| dB
1	0	1	0
10	1	1	0
10^2	2	1	0
10^3	3	1	0
10^4	4	1/5	20 log(1/5) = –13.98
10^5	5	1/50	20 log(1/50) = –33.98
10^6	6	1/500	20 log(1/500) = –53.98
para#2 R = 1Ω; L = 1mH; L/R = 1 ms			
ω (rad/s)	$\log_{10}(\omega)$	|G(jω)|	20 logG(jω)| dB
1	0	1	0
10	1	1	0
10^2	2	1	0
10^3	3	1	0
10^4	4	1/10	20 log(0.1) = –20
10^5	5	1/100	20 log(0.01) = –40
10^6	6	1/1000	20 log(0.001)= –60
para#3 R = 1Ω; L = 2 mH; L/R = 2 ms			
ω (rad/s)	$\log_{10}(\omega)$	|G(jω)|	20 logG(jω)| dB
1	0	1	0
10	1	1	0
10^2	2	1	0
10^3	3	1/2	20 log(1/2) = –6.02
10^4	4	1/20	–26.02
10^5	5	1/200	–46.02
10^6	6	1/2000	–66.02

4.2.3 Gain margin and phase margin

Consider the closed loop negative feedback system shown in Figure 4.4, let

$G_p(s)$ – open loop gain of the plant
$P_{do}(s)$ – desired output of the plant
$P_{ao}(s)$ – actual output of the plant

One may note that the transfer function of the closed loop system $G_{pcl}(s) = \dfrac{G_p(s)}{1+G_p(s)}$; the gain of the closed system would be infinity if the characteristic equation $1 + G_p(s)$ is equal to zero. Meaning that the condition,

Table 4.2 Phase Angles of Gains of R-L Network for Different Parameter Values

para#1 R = 1Ω; L = 0.5mH; L/R = 0.5 ms		
ω (rad/s)	$\log_{10}(\omega)$	$-\tan^{-1}(\omega L/R)$
1	0	–0.0286
10	1	–0.2865
10^2	2	–2.8624
10^3	3	–26.561
10^4	4	–78.699
10^5	5	–88.8542
10^6	6	–89.885
para#2 R = 1Ω; L = 1mH; L/R = 1 ms		
ω (rad/s)	$\log_{10}(\omega)$	$-\tan^{-1}(\omega L/R)$
1	0	–0.0573
10	1	–0.573
10^2	2	–5.7106
10^3	3	–45
10^4	4	–84.3
10^5	5	–89.427
10^6	6	–89.9427
para#3 R = 1Ω; L = 2 mH; L/R = 2 ms		
ω (rad/s)	$\log_{10}(\omega)$	$-\tan^{-1}(\omega L/R)$
1	0	–0.1145
10	1	–1.1458
10^2	2	–11.3099
10^3	3	–63.4349
10^4	4	–87.1376
10^5	5	–89.7135
10^6	6	–89.97

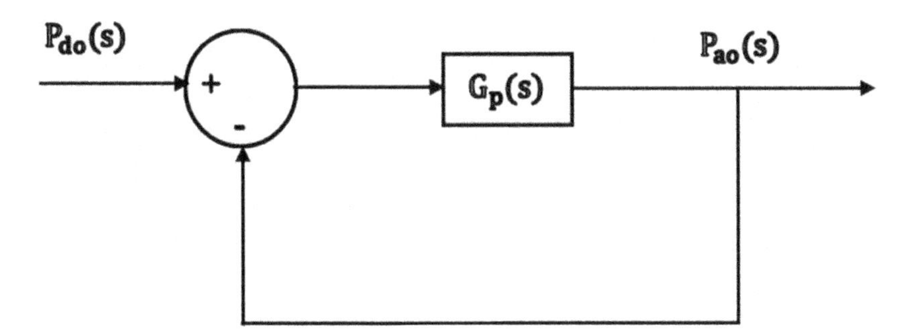

Figure 4.4 A negative feedback closed loop control operation.

open loop transfer function $G_p(j\omega) = -1$ would result an instability in the closed loop system. A gain of –1 implies that magnitude of the output is equal to the input with a phase difference of –180°; or in other words as the frequency of the sinusoidal input is varied for a certain frequency the output shall be an inverted input and eventually leads to an unstable operation. On a Bode plot the magnitude of –1 is 0 dB with a phase difference of –180°.

Gain margin (Gm) of a system indicates how far is the gain of the system from the 0 dB line at phase cross over frequency (ω_{pc}) where the phase plot crosses the phase angle –180°; the gain of a system can be increased/decreased by a factor k without reaching instability and is related to Gm in dB as follows:

$$G_m = 0\text{db} - 20\log_{10}\left|G_{(j\omega pc)}\right| = 20\log_{10}\frac{1}{\left|G\left(j\omega_{pc}\right)\right|} = 20\log_{10}k \tag{4.10}$$

The phase margin (Pm) of a system is a measure of the phase angle of the system at gain cross over frequency (ω_{gc}) where the gain crosses the 0 dB line and can be calculated as follows:

$$Pm = 180 + \angle G\left(j\omega_{gc}\right)$$

For the closed loop system to be stable both Gm and Pm of the open loop system should be positive; positive margins indicate still there is a safety margin before instability. Negative margins predict instability in a closed loop system; however choosing a lower gain of maximum value of k a stable closed system can be achieved.

Consider a system having the transfer function

$$G(s) = \frac{-334s + 36000}{s^2 + 13.3s + 1440} \tag{4.12}$$

The MATLAB command, margin(tf([-334 36000],[1 13.3 1440])) could be used to plot the Bode plot with gain margin and phase margin shown in Figure 4.5; one may note that negative values of Gm and Pm predict instability in the closed loop system.

Pole-zero map in Figure 4.6 of the above open loop system shows that there are complex conjugate poles on the left half of the s-plane and there is a zero on the right half of the s-plane; one may note that step response in Figure 4.7 of the open loop system is stable. One may use the following MATLAB code to plot the pole-zero map and step response of the open loop system given by Eqn. 4.12.

```
>> np = [–334 36000];
>> dp = [1 13.3 1400];
>> sys1 = tf(np,dp)
```

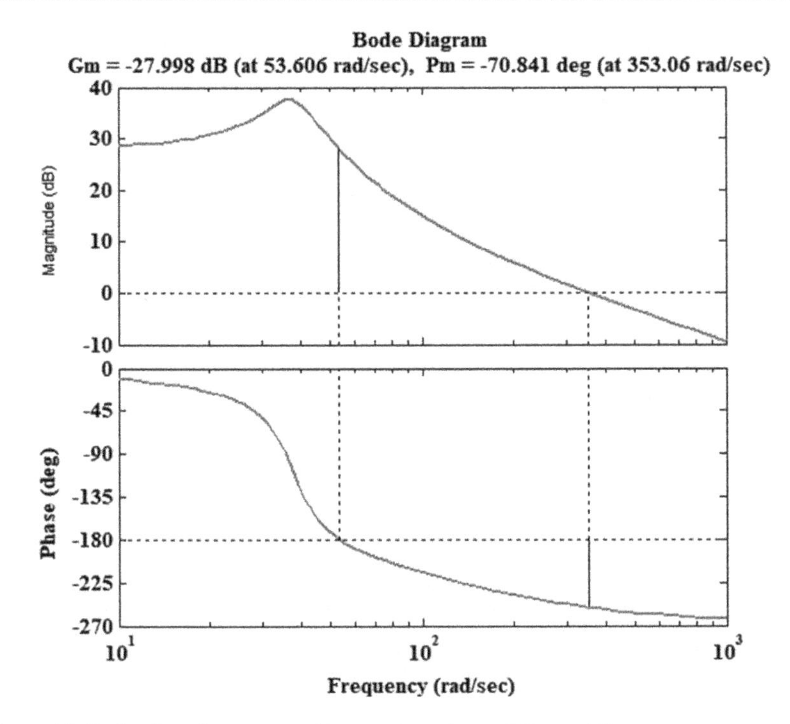

Figure 4.5 Bode plot of a system with negative values of Gm and Pm.

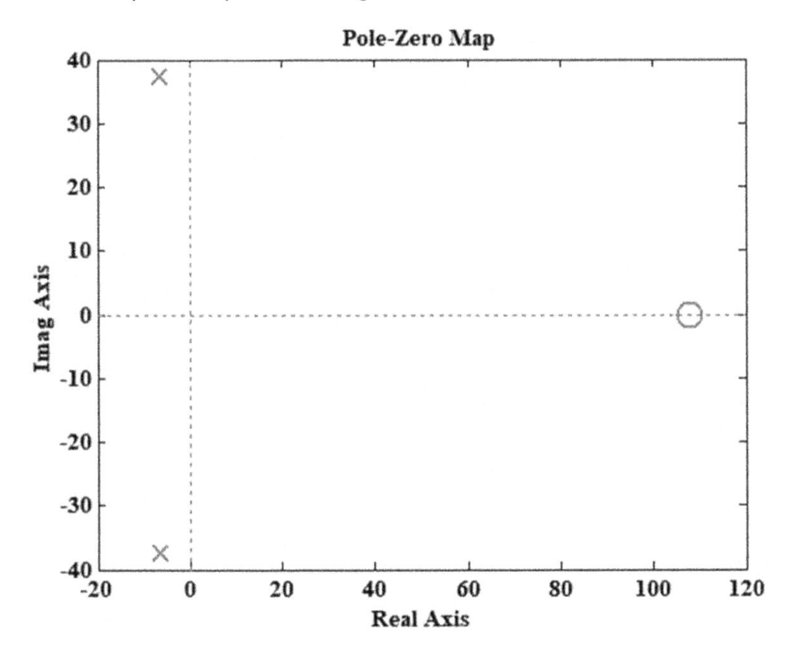

Figure 4.6 Pole-zero map of an open loop system with a zero on the right half of the s-plane.

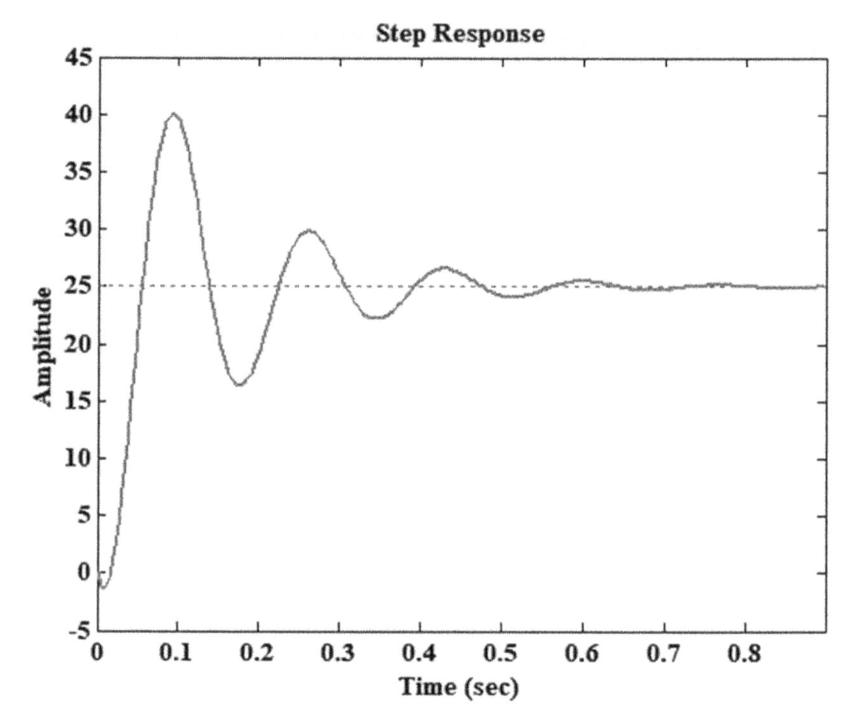

Figure 4.7 Step response of an open loop system with a zero on the right half of the s-plane.

```
Transfer function:
-334 s + 36000
------------------
s^2 + 13.3 s + 1400
>> figure (1)
>> pzmap (sys1)
>> figure (2)
>> step(sys1)
```

However, a pole-zero map of the closed loop system shown in Figure 4.8 shows that there are complex conjugate poles and a zero on the right hand side of the s-plane making the closed loop system unstable. The MATLAB code to plot the pole-zero map of the closed loop system is as follows:

```
>> np = [–334 36000];
>> dp = [1 13.3 1400];
>> [ncl,dcl] = feedback(np,dp,1,1);
>> sys = tf(ncl,dcl)
Transfer function:
```

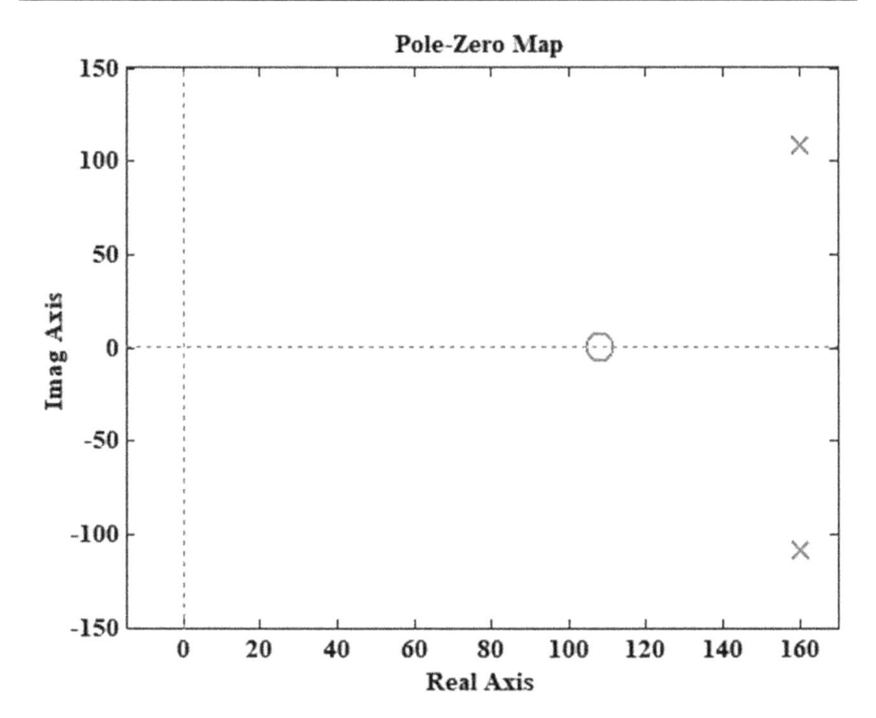

Figure 4.8 Pole-zero map of the closed loop system.

```
–334 s + 36000
---------------------
s^2–320.7 s + 37400
>> pzmap(sys)
```

Thus, since both Gm and Pm of the open loop system is negative as depicted in Figure 4.5, the closed loop system is unstable as illustrated in Figure 4.9. One may use the following MATLAB commands to plot the step response shown in Figure 4.9.

```
>> np = [–334 36000];
>> dp = [1 13.3 1400];
>> [ncl,dcl] = feedback(np,dp,1,1);
>> step(tf(ncl,dcl))
```

4.2.4 Impacts of controller gain on stability

One may note that for the above system,

$$G_m = -27.998 = 20\log_{10} k \tag{4.13}$$

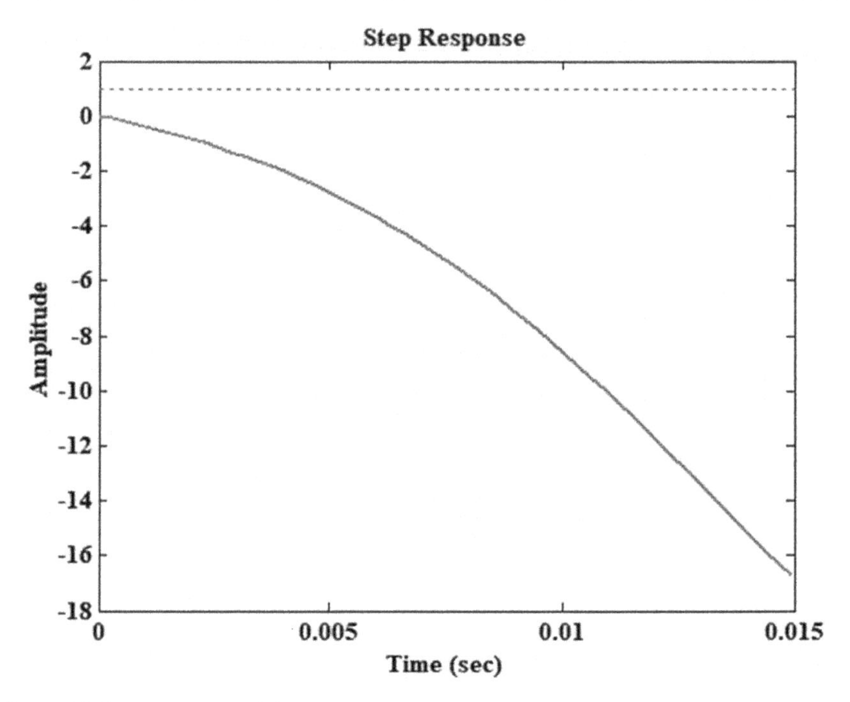

Figure 4.9 Step response of the system.

and the factor k can be calculated as follows:

$$-1.3999 = \log_{10}(k) \tag{4.14}$$

$$-1.3999 = \log_{10}\left(10^{-1.3999}\right) \tag{4.15}$$

$$k = 10^{-1.3999} = 0.0399 \tag{4.16}$$

The above system can be brought from instability to stability by decreasing the gain to a maximum value of 0.039 as shown in Figure 4.10. Following MATLAB codes are used to introduce the gain factor to obtain the Bode plot shown in Figure 4.11; one may note that both Gm and Pm are positive after introducing a gain factor of 0.039.

```
>> np = [–334 36000];
  >> dp = [1 13.3 1400];
  >> sys1 = tf(np,dp)
  Transfer function:
  –334 s + 36000
  --------------------
```

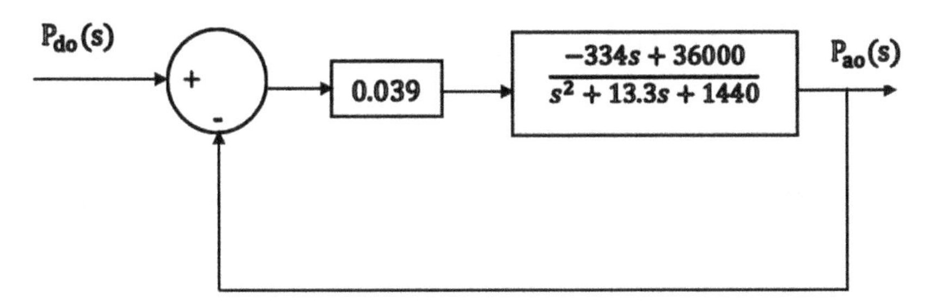

Figure 4.10 Closed loop operation of a plant with a gain of 0.039.

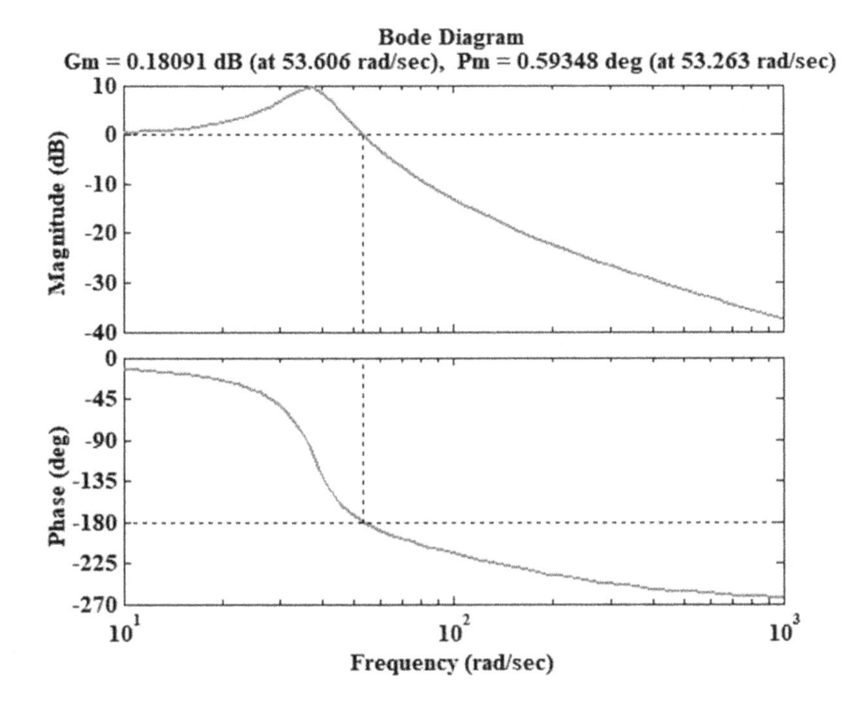

Figure 4.11 Bode plot of the open loop system for a gain of 0.039.

```
s^2 + 13.3 s + 1400
>> margin(sys1*0.039)
```

The closed loop response of the system with a gain factor of 0.039 is plotted in Figure 4.12 using the following MATLAB commands; one may note that the new closed system is marginally stable.

```
>> np = [−334 36000];
  >> d p = [1 13.3 1400];
```

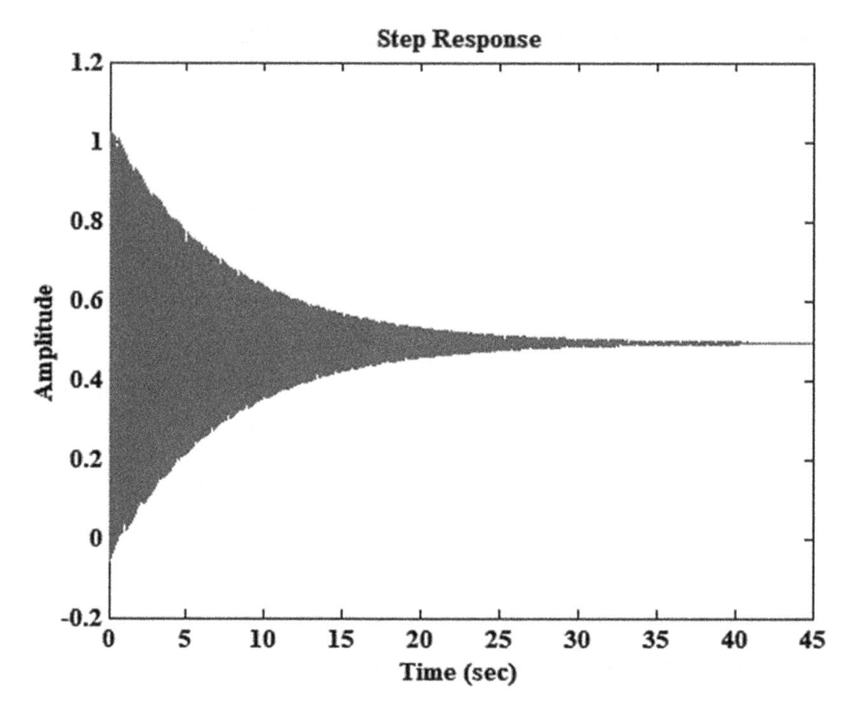

Figure 4.12 Step response of the closed loop system for a gain of 0.039.

```
>> sys1 = tf(np,dp);
>> format long G
>> sys1*0.039
Transfer function:
-13.03 s + 1404
-------------------
s^2 + 13.3 s + 1400
>> npk = [-13.03 1404];
>> dpk = dp;
>> [nclk,dclk] = feedback(npk,dpk,1,1);
>> step(tf(nclk,dclk))
```

In Figure 4.13, the Bode plots of the above open loop system with unity gain, with a gain 0.01 less than k and with a gain 0.05 greater than k are plotted using the following MATLAB codes. One note may that as the gain changes the magnitude plot varies but the phase plot remains same for all; as the gain cross over frequency changes Pm changes.

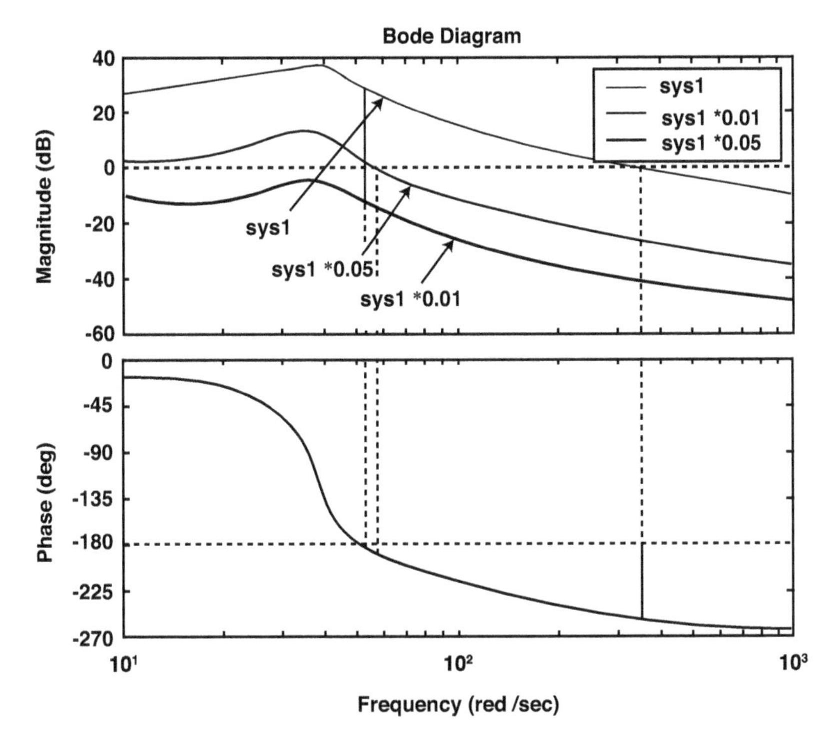

Figure 4.13 Bode plot of a system for different gains.

```
>> np = [–334 36000];
>> dp = [1 13.3 1400];
>> sys1 = tf(np,dp);
>> margin(sys1)
>> hold on
>> margin(sys1*0.01)
>> margin(sys1*0.05)
```

With unity gain, both Gm and Pm are negative as follows:

Gm = –27.998 dB (at 53.606 rad/sec), Pm = –70.841 deg (at 353.06 rad/sec).

However, with a gain of 0.01 less than the gain factor k both Gm and Pm are positive as follows:

Gm = 12.002 dB (at 53.606 rad/sec), Pm = Inf.

And, with a gain of 0.05 slightly greater than the factor k both Gm and Pm become negative as follows:

Gm = –1.9772 dB (at 53.606 rad/sec), Pm = –6.0885 deg (at 57.726 rad/sec).

4.3 ROOT LOCUS

Root locus technique is used to choose the gain of a closed loop system to achieve the desired output performance. Consider the R-C circuit in Figure 2.11, with R = 100Ω; C = 100μF, the open loop transfer function

$$\frac{V_c(s)}{V_s(s)} = G_p(s) = \frac{100}{s+100} \quad (4.17)$$

One may note that transfer function of the closed loop system of the R-C circuit with unity negative feedback and with a controller of gain k shown in Figure 4.14 could be written as

$$G_{pcl}(s) = \frac{k\left(\frac{100}{s+100}\right)}{1+k\left(\frac{100}{s+100}\right)} \quad (4.18)$$

Root locus is plotted using the roots of the characteristic equation

$$1 + kG_p(S) = 1 + k\frac{100}{s+100} = 0 \quad (4.19)$$

for different values of k; meaning that roots of the equation,

$$s + 100 + 100k = 0 \quad (4.20)$$

against various gains.

The values of s for different values of k are calculated as seen in Table 4.3.

Once the root locus is plotted, different values of gain k could be chosen to check the output for the final selection of the value of k.

One may use the MATLAB program given in mcode4–2 to plot the root locus and to select the gain of the controller. The values of k for the four different points on the root locus shown in Figure 4.15 chosen are k1 = 0.0242;

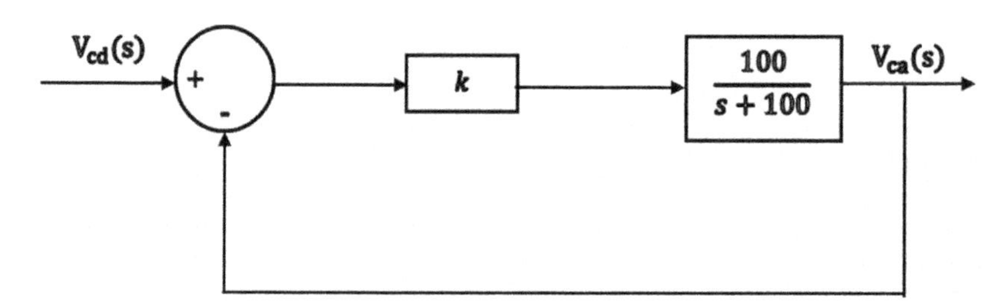

Figure 4.14 A closed loop system of the R-C circuit.

Table 4.3 Roots of the Characteristic Equation for Different Values of Gains

k	s
0	–100
1	–200
5	–600
10	–1100
100	–10100
1000	–100100
∞	–∞

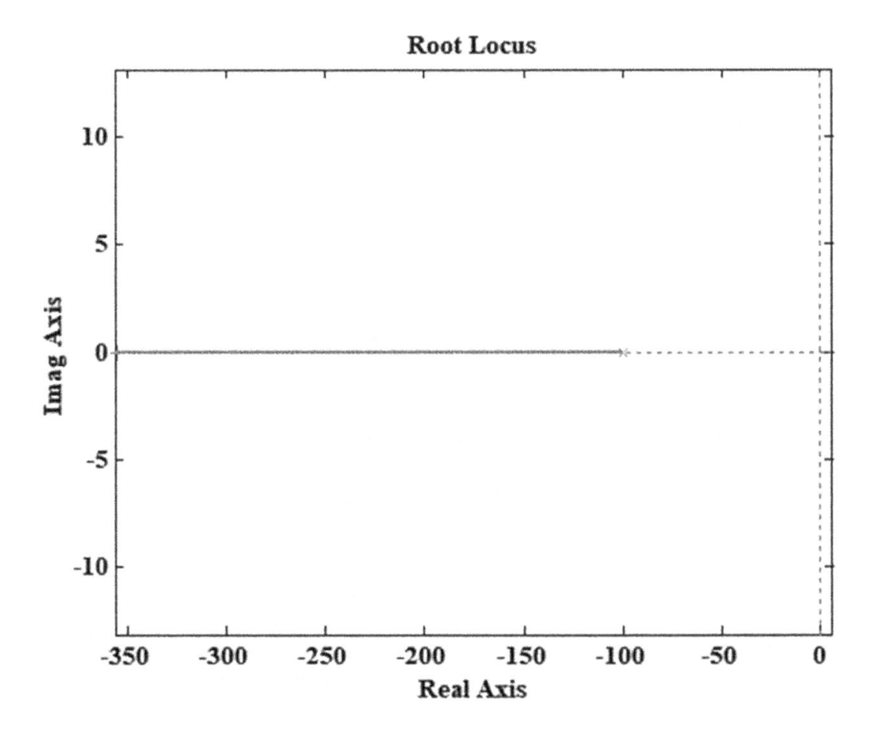

Figure 4.15 Root locus of an R-C circuit.

k2 = 1.1924; k3 = 2.06 and k4 = 2.5496. The step responses for the selected values are shown in Figure 4.16. One may note that, with low values of controller gain, the amplitude is very small compared with higher values. It should also be noted that output response reaches steady state faster with high values of controller gain as compared to low values of gain.

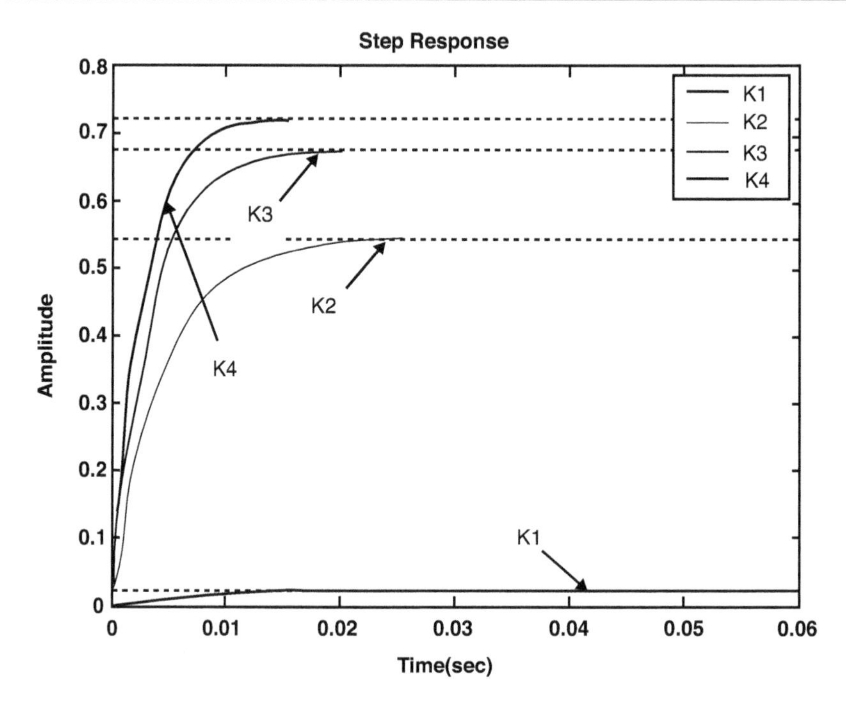

Figure 4.16 Step response of the RC closed loop system for different values of k.

mcode4–2

```
% A program to compare the output performance of an R-C circuit
  % Different values of gain are selected from the root locus
  % Gp = 100/(s + 100)
  % k is initialized to 1 and unity feedback is assumed
  np = [100]; dp = [1 100]; nc = [1]; dc = [1]; nh = [1]; dh = [1];
  Count = 1;
  while count<5
  figure (1)
  rlocus(np,dp)
  [k,p] = rlocfind(np,dp)
  nlp = conv(np,k*nc);
  dlp = conv(dp,dc);
  [nclp,dclp] = feedback(nlp,dlp,nh,dh);
  sys = tf(nclp,dclp);
  figure (2)
  hold on
  step(sys);
  count = count + 1;
  end
```

BIBLIOGRAPHY

Benjamin C. K., Automatic control systems (5th ed.). India: Prentice Hall of India Private Limited, 1989.

Dorf R. C. and Bishop R. H., Modern control systems (1st ed.). World student series. USA: Addison-Wesley, 1998.

Electrical systems/basic alternating (AC) theory/transfer function analysis, online http://control.com/textbook/ac-electricity/transfer-function-analysis/, 2022.

Hadi S., Power system analysis (2nd ed.). Singapore: McGraw-Hill Education (Asia), 2004.

Manke B. S., Linear control systems with MATLAB applications (8th ed.). India: Khanna Publishers, 2005.

Nagoor K. A., Control systems (1st ed.). India: R B A Publications, 1998.

Ogata K., Modern control engineering (3rd ed., International). USA: Prentice Hall International Inc., 1997.

Part-Enander E., Sjoberg A., Melin B., and Ishaksson P., The MATLAB handbook (1st ed.). USA: Addison Wesley, 1996.

Umanand L., Switched mode power conversion, online http://nptel.ac.in/, 2014.

Chapter 5

Operation of a buck converter circuit

5.1 INTRODUCTION

Consider the PSB model of the open loop buck converter circuit in Figure 1.20 with system specifications as seen in Table 5.1. The buck converter circuit mainly consists of a power circuit and a control circuit; the power circuit consists of an input power supply, a controllable switch S_{1bu}, a free-wheeling diode S_{2bu}, an inductor, a capacitor and a load; one may use GTO or MOSFET as the controllable switch. The control circuit controls the turn ON and turn OFF operation of the switches. A pulse generator circuit available in the PSB library is used to generate the switching pulses in Figure 5.1 to control the operation of the switches.

One may note that the switching period,

$$T_s = T_{on} + T_{off} \tag{5.1}$$

the switching frequency,

$$f_{bu} = \frac{1}{T_s} \tag{5.2}$$

the duty cycle,

$$d_{bu} = \frac{T_{on}}{T_s} = \frac{T_{on}}{T_{on} + T_{off}} \tag{5.3}$$

and for a buck converter,

$$d_{bu} = \frac{V_{cbu}}{V_{sbu}} \tag{5.4}$$

5.2 OPERATION OF THE POWER CIRCUIT

The switching pulses (sp) generated from the control circuit is used to turn ON and turn OFF the controllable switch for the successful operation of the buck converter circuit. Operation of the power circuit can be divided into mode 1 and mode 2.

DOI: 10.1201/9781003511236-5

Table 5.1 System Specifications of the PSB Model of the Buck Converter

Input	*Inductance*	*Capacitance*	*Load resistance*	*Switching pulses*		
V_s (V)	L_{bu} (mH)	C_{bu} (μF)	R_{bu} (Ω)	Period T (s)	Duty cycle D	Amplitude (V)
9	10	100	5	1 e^{-3}	55.56 %	1

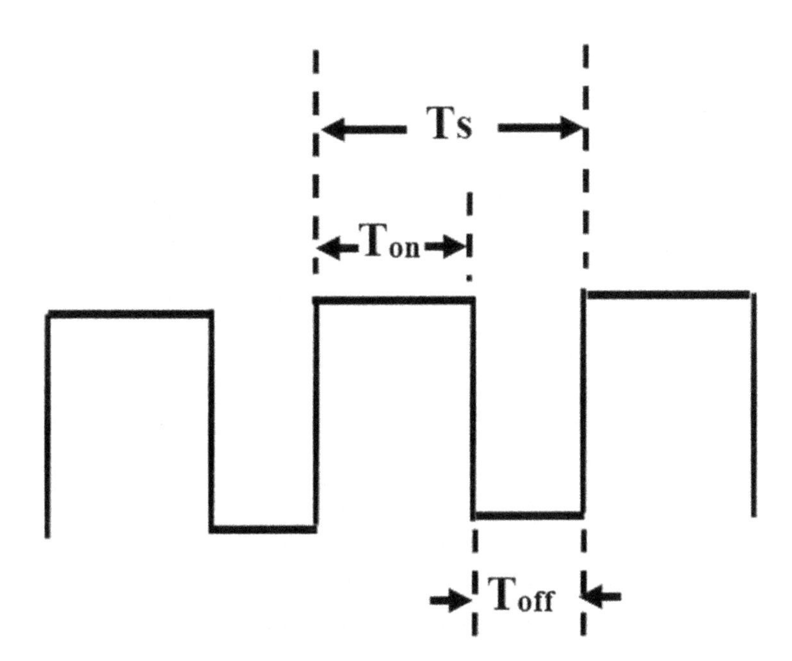

Figure 5.1 Switching pulses.

5.2.1 Operation of the power circuit in mode 1

In mode 1, when sp is equal to one, the switch S_{1bu} would be in the ON position and the source V_{sbu} is connected to the inductor and capacitor through the switch S_{1bu} as shown in Figure 5.2; one may note that as the free-wheeling diode S_{2bu} is connected across V_{sbu} with a reverse polarity, S_{2bu} would be in the OFF position. The inductor current increases and the capacitor gets charged as illustrated in Figure 5.7; one may note that in mode 1 the voltage across the inductor (v_{indbu}), the inductor current (i_{indbu}) and the voltage across the capacitor (v_{cbu}) are positive. However, in mode 1 as the inductor current increases, the capacitor current (i_{cbu}) reverses the polarity and satisfy Kirchoff's current law (KCL) at the node depicted in Figure 5.3.

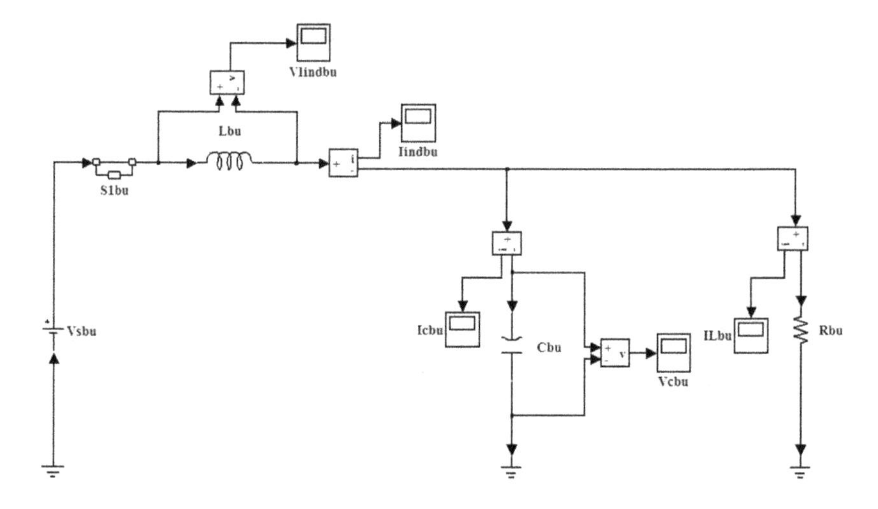

Figure 5.2 Power circuit of the buck converter in mode 1.

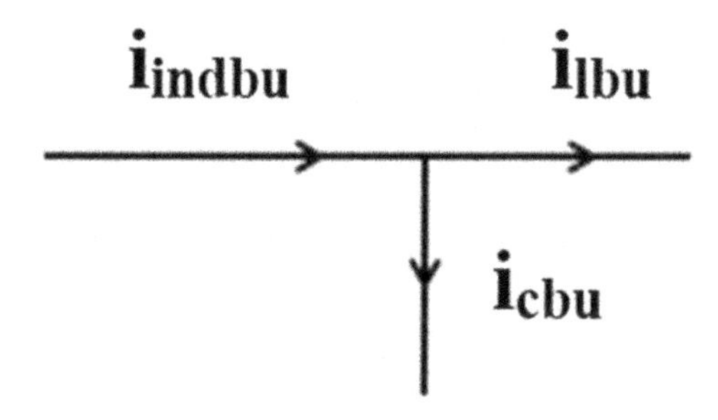

Figure 5.3 Assumed directions of currents following the ammeter connections.

5.2.1.1 Retrieving instantaneous values of current in mode 1

Following the ammeter connections in Figure 5.2, one may apply KCL at the node of Figure 5.3 to verify the values of currents obtained.

One may get,

$$\overrightarrow{i_{indbu}} = \overrightarrow{i_{cbu}} + \overrightarrow{i_{lbu}}$$

Retrieving the instantaneous values i_{indbu}, i_{cbu} and i_{lbu} at t = 0.132 s of Figure 5.8 using the MATLAB® command ginput(1): one may get,

```
>>[t,iindbu] = ginput(1)
  t =
  0.1320
  iindbu =
  0.8986
```

```
>>[t,icbu] = ginput(1)
t =
0.1320
icbu =
-0.0932
>>[t,ilbu] = ginput(1)
t =
0.1320
ilbu =
0.9936
```

Substituting the values in Eqn. 5.5, one may get,

$$0.8986 \approx -0.0932 + 0.9936$$

5.2.1.2 Retrieving instantaneous values of voltages in mode 1

Following the voltmeter connections in Figure 5.2, one may apply KVL to the loop of Figure 5.4 to verify the values of voltages obtained.

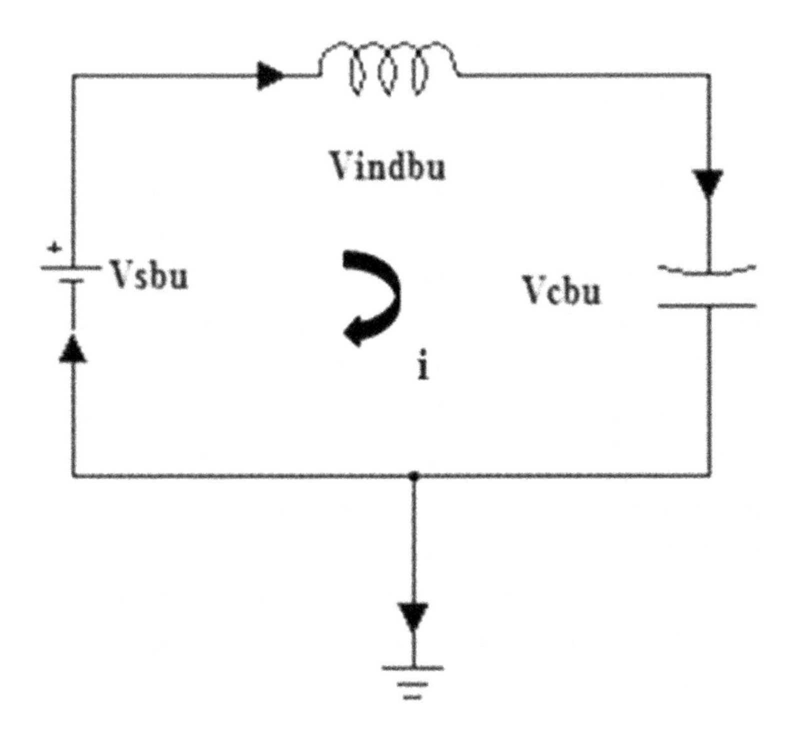

Figure 5.4 Voltage loop in mode 1.

One may get,

$$V_{sbu} - V_{indbu} - V_{cbu} = 0$$

$$V_{sbu} = V_{indbu} + V_{cbu}$$

Retrieving the instantaneous values V_{sbu}, v_{indbu} and v_{cbu} at t = 0.132 s of Figure 5.9 using the MATLAB command ginput(1): one may get,

```
>>[t,Vsbu] = ginput(1)
t =
0.1320
Vsbu =
8.9966
>>[t,vindbu] = ginput(1)
t =
0.1320
vindbu =
4.0610
>>[t,vcbu] = ginput(1)
t =
0.1320
vcbu =
4.8203
```

Substituting the values in Eqn. 5.7 and neglecting the voltage drop across S_{1bu} one may get,

$$8.9966 \approx 4.0610 + 4.8203$$

5.2.2 Operation of the power circuit in mode 2

In mode 2, the switch S_{1bu} would be in the OFF position and the source V_{sbu} would be disconnected as shown in Figure 5.5; since the current through the inductor cannot be interrupted instantaneously, the voltage across the inductor reverses polarity to forward bias the free-wheeling diode S_{2bu} and S_{2bu} would be in the ON position as shown in Figure 5.5.

Analysis of Figure 5.7 shows that inductor current starts decreasing and polarity of the capacitor current changes to meet the load requirement without any changes in the polarity of the capacitor voltage.

5.2.2.1 Retrieving instantaneous values of currents in mode 2

To verify the values of currents obtained, one may apply KCL at the node of Figure 5.3. Retrieving the instantaneous values i_{indbu}, i_{cbu} and i_{lbu} at

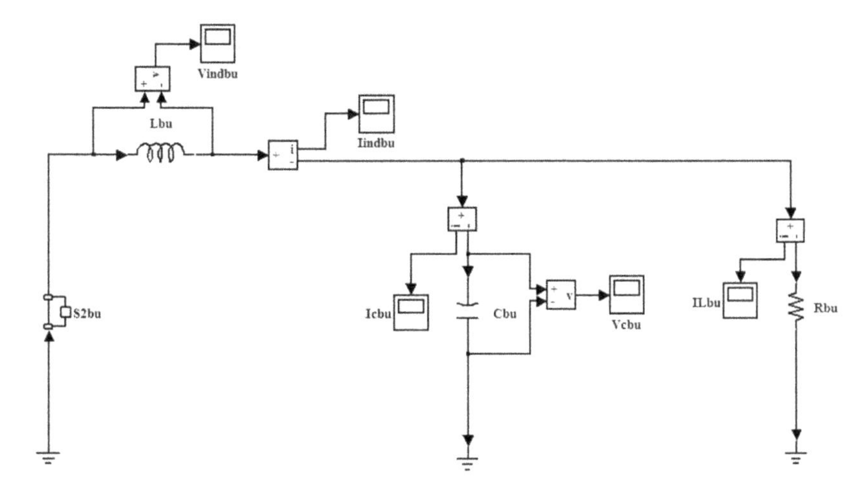

Figure 5.5 Power circuit of the buck converter in mode 2.

t = 0.1326 s of Figure 5.8 using the MATLAB command ginput(1): one may get,

```
>>[t,iindbu] = ginput(1)
 t =
 0.1326
 iindbu =
 1.0837
 >>[t,icbu] = ginput(1)
 t =
 0.1326
 icbu =
 0.0729
 >>[t,ilbu] = ginput(1)
 t =
 0.1326
 ilbu =
 1.0031
```

Substituting the values in Eqn. 5.5, one may get,

$$1.0837 \approx 0.0729 + 1.0031$$

5.2.2.2 Retrieving instantaneous values of voltages in mode 2

Following the voltmeter connections in Figure 5.5, one may apply KVL to the loop of Figure 5.6 to verify the values of voltages obtained.

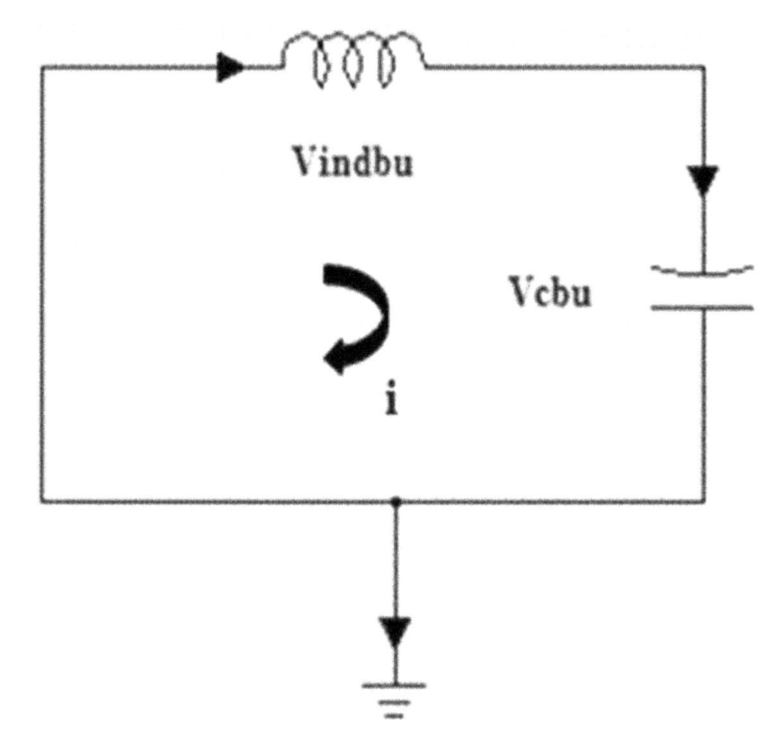

Figure 5.6 Voltage loop in mode 2.

One may get,

$$-v_{indbu} - v_{cbu} = 0$$

$$-v_{indbu} = v_{cbu}$$

Retrieving the instantaneous values v_{indbu} and v_{cbu} at t = 0.1325 s of Figure 5.9 using the MATLAB command ginput(1): one may get,

```
>>[t,vindbu] = ginput(1)
  t =
  0.1326
  vindbu =
  -5.1051
  >>[t,vcbu] = ginput(1)
  t =
  0.1326
  vcbu =
  5.0373
```

Substituting the values in Eqn. 5.9 and neglecting the voltage drop across S_{2bu}, one may get,

$$-(-5.1051) \approx 5.0373$$

Note: One may note that as the voltage across the inductor reverses polarity, v_{indbu} is negative.

5.3 PERFORMANCE ANALYSIS OF A BUCK CONVERTER CIRCUIT WITH PARAMETER VARIATIONS

One may analyze the performance of the open loop buck converter in Figure 1.20, for different switching frequencies, for different values of the peak-peak ripple current of the inductor and for different values of peal-peak ripple voltage of the capacitor as follows:

Let

V_{sbu}—Input voltage (V); I_{Lbu}—load current (A); f_{bu}—Switching frequency (Hz);
I_{indbu}—Inductor current (A); L_{bu}—inductance (H); C_{bu}—Capacitance (F); d_{bu}_duty cycle; ΔI_{indbu}—Peak-peak ripple current of the inductor (A); V_{cbu}—Capacitor voltage (V); and
ΔV_{cbu}—Peak-peak ripple voltage of the capacitor (V).

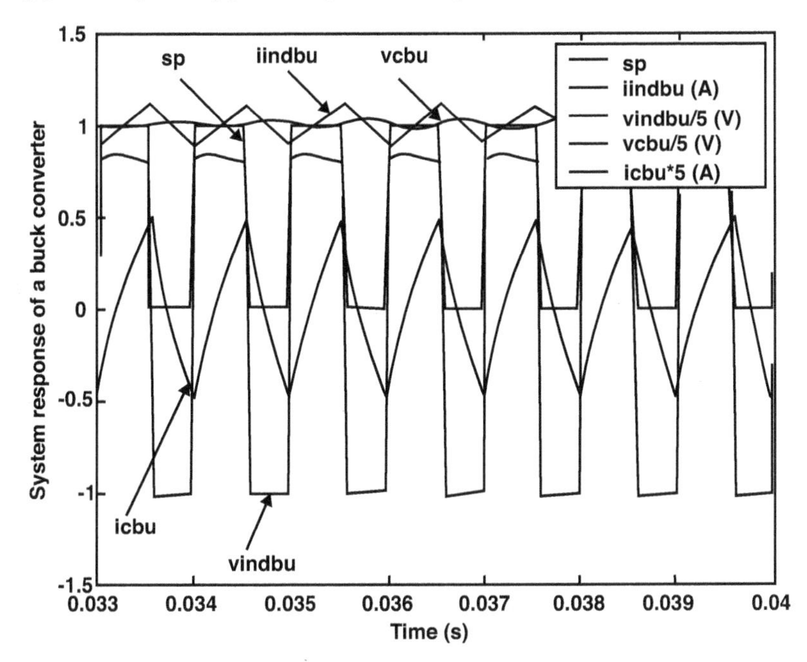

Figure 5.7 Switching characteristics of the inductor and capacitor in a buck converter.

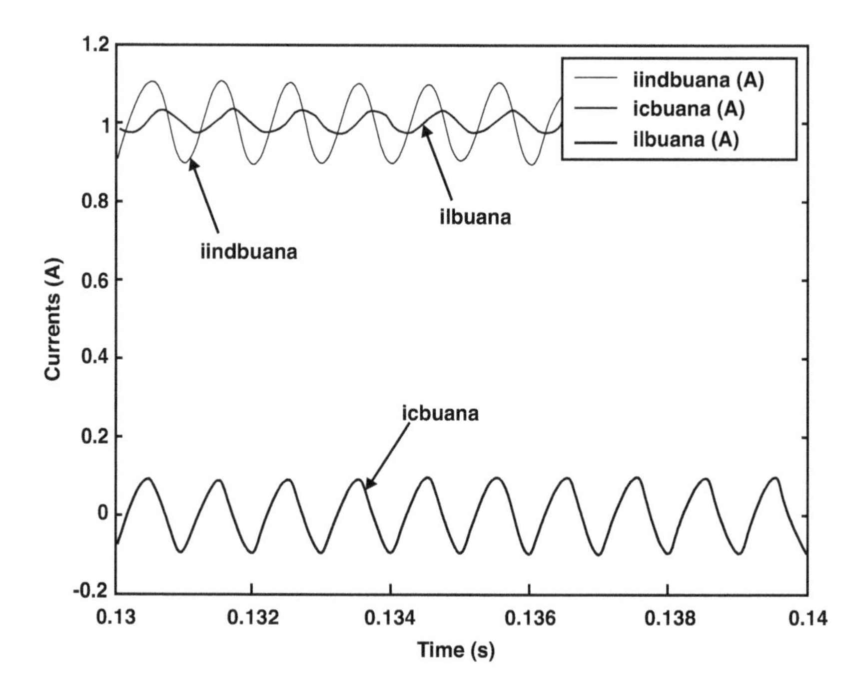

Figure 5.8 Instantaneous values of currents for analysis.

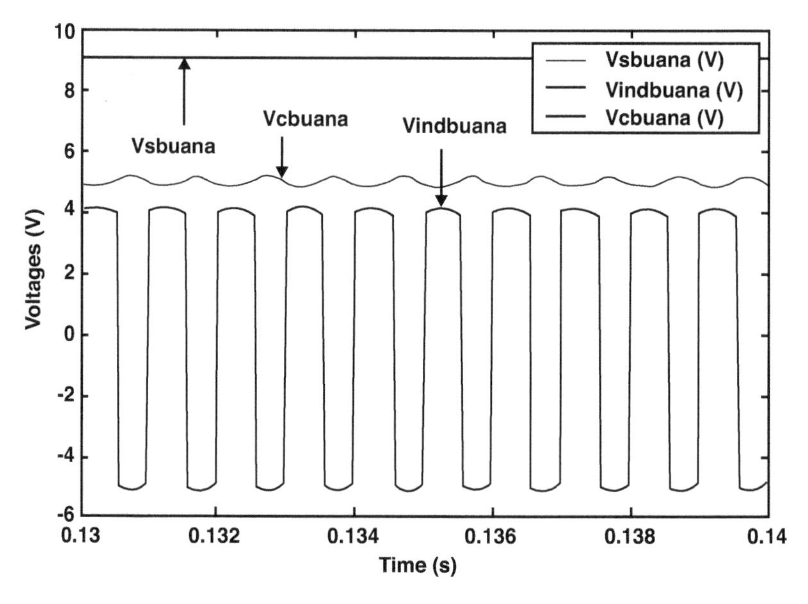

Figure 5.9 Instantaneous values of voltages for analysis.

For a buck converter, the peak-peak ripple current of the inductor is given by

$$\Delta I_{indbu} = \frac{1}{L_{bu} f_{bu}} V_{sbu} D_{bu} \left(1 - D_{bu}\right)$$

And peak-peak ripple voltage of the capacitor is given by

$$\Delta V_{cbu} = \frac{V_{sbu} D_{bu} \left(1 - D_{bu}\right)}{8 L_{bu} C_{bu} f_{bu}^2}$$

The values of inductance and capacitance obtained for various frequencies and for different specifications are calculated as seen in Table 5.2–Table 5.5. Analysis of Figure 5.10 to

Figure 5.13 shows that for low frequencies capacitor is unable to build up voltage; however with higher frequencies capacitor is able to build up more

Table 5.2 Values of Inductance and Capacitance for Different Frequencies as per Specification # 1

Parameter labels	f_{bu} *(Hz)*	L_{bu} *(H)*	C_{bu} *(F)*
parabu#11	1	22.22	0.125
parabu#12	10	2.222	0.0125
parabu#13	100	222 m	1250μ
parabu#14	1k	22.2 m	125μ
parabu#15	10k	2.22 m	12.5μ

Table 5.3 Values of Inductance and Capacitance for Different Frequencies as per Specification # 2

Parameter labels	f_{bu} *(Hz)*	L_{bu} *(H)*	C_{bu} *(F)*
parabu#21	1	11.11	0.125
parabu#22	10	1.11	0.0125
parabu#23	100	111 m	1250μ
parabu#24	1k	11.1 m	125μ
parabu#25	10k	1.1 m	12.5μ

Table 5.4 Values of Inductance and Capacitance for Different Frequencies as per Specification # 3

Parameter labels	f_{bu} *(Hz)*	L_{bu} *(H)*	C_{bu} *(F)*
parabu#31	1	7.41	0.125
parabu#32	10	741 m	0.0125
parabu#33	100	74.1 m	1250μ
parabu#34	1k	7.41 m	125 μ
parabu#35	10k	0.741 m	12.5μ

Table 5.5 Values of Inductance and Capacitance for Different Frequencies as per Specification # 4

Parameter labels	f_{bu} *(Hz)*	L_{bu} *(H)*	C_{bu} *(F)*
parabu#41	1	111.11	0.8342
parabu#42	10	11.11	0.08342
parabu#43	100	1.11	8342µ
parabu#44	1k	110 m	834.2µ
parabu#45	10k	11.1 m	83.42µ

and more voltage. It should be noted that the performance of an open loop buck converter varies with parameter variations as illustrated in Figure 5.10–Figure 5.13. To have an insight into the operation of a buck converter, parabu#44 has randomly chosen and the voltage across the switching elements and energy storing elements are illustrated in Figure 5.14; one may note that at a time, one switching element only conducts, the voltage across the inductor reverses polarity and the voltage across the capacitor never reverses the polarity.

Specification # 1

V_{sbu} = 9 V; V_{cbu} = 5 V; D_{bu} = 5/9; I_{Lbu} = 1 A; ΔI_{indbu} = 100 mA (10 % of the full load current); ΔV_{cbu}=100 mV (2 % of the output voltage).

5.3.1 Calculation for L_{bu} and C_{bu} for f_{bu}=1 Hz

Substituting the values in Eqn. 5.10, one may get,

$$100 \times 10^{-3} = \frac{9 \times \frac{5}{9} \times \frac{4}{9}}{1 \times L_{bu}}$$

$$\therefore L_{bu} = 22.22\,\text{H}$$

Substituting the values in Eqn. 5.11, one may get,

$$100 \times 10^{-3} = \frac{9 \times \frac{5}{9} \times \frac{4}{9}}{8 \times 22.22 \times C_{bu} \times 1^2}$$

$$\therefore C_{bu} = 0.125\,\text{F}$$

Specification # 2

V_{sbu} = 9 V; V_{cbu} = 5 V; D_{bu} = 5/9; I_{Lbu}=1 A; ΔI_{indbu} = 200 mA (20 % of the full load current); ΔV_{cbu} = 200 mV (4 % of the output voltage).

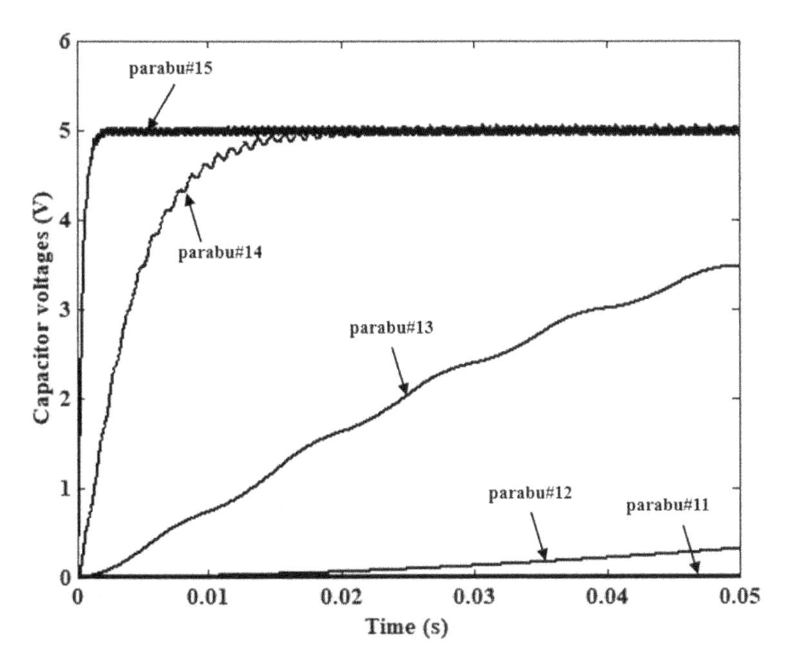

Figure 5.10 Plots of capacitor voltages for different frequencies as per specification # 1.

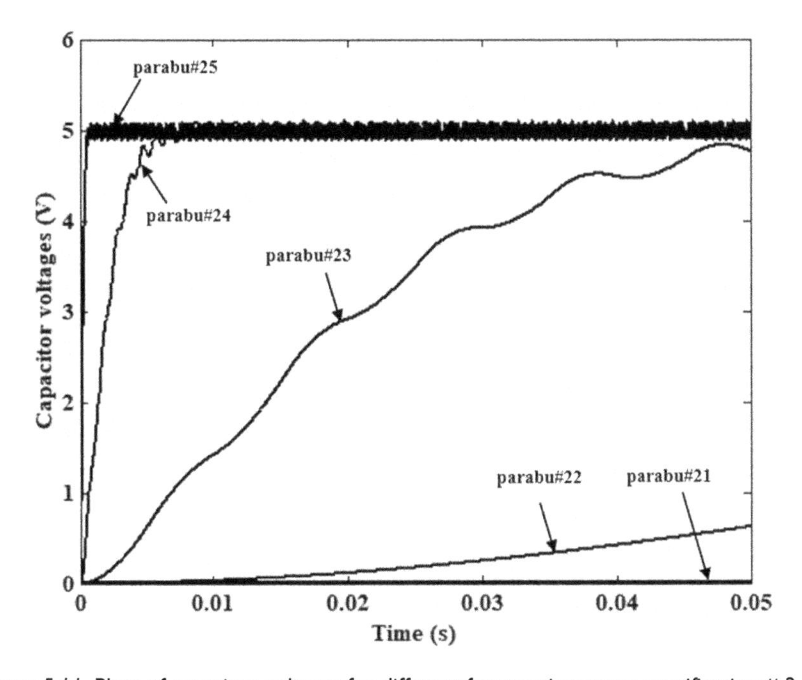

Figure 5.11 Plots of capacitor voltages for different frequencies as per specification # 2.

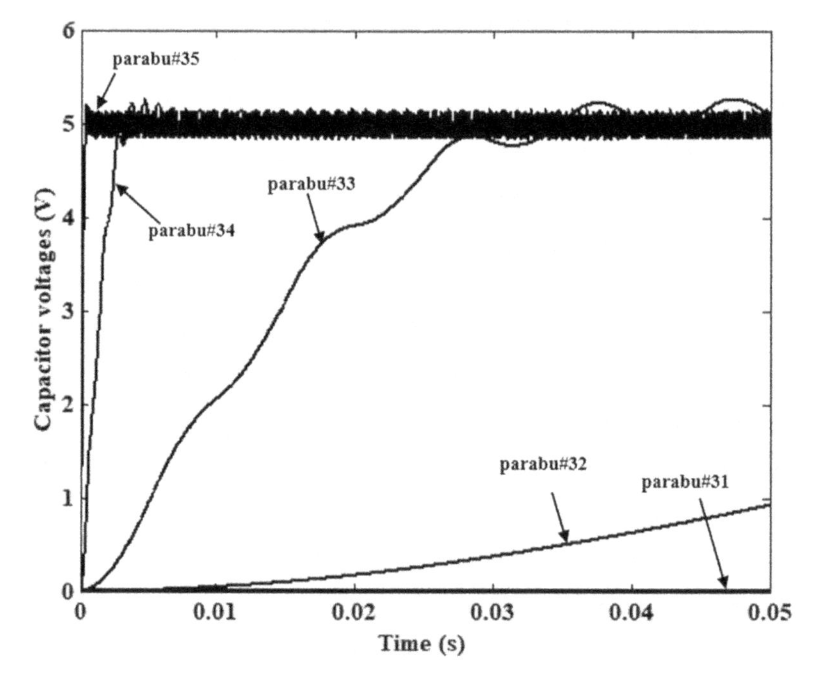

Figure 5.12 Plots of capacitor voltage for different frequencies as per specification # 3.

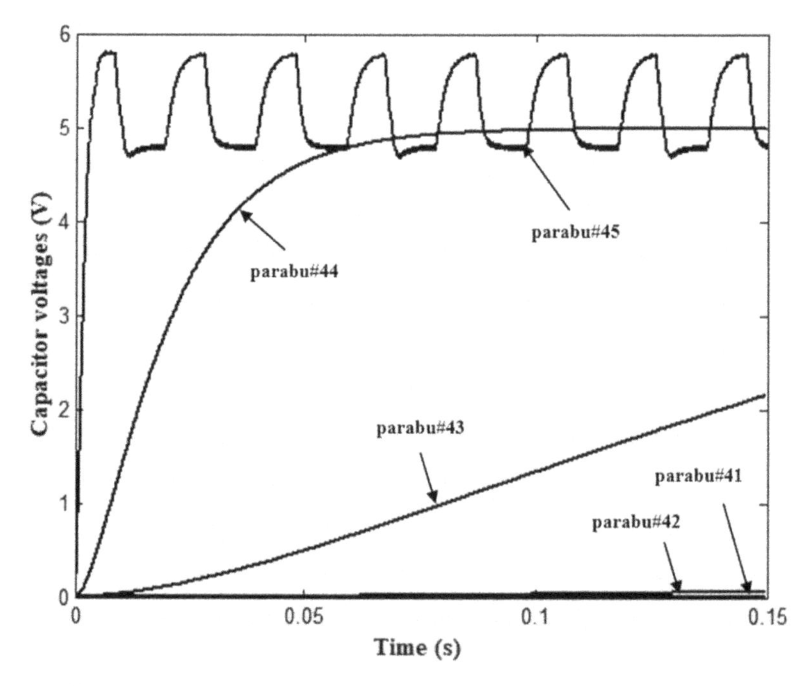

Figure 5.13 Plots of capacitor voltage for different frequencies as per specification # 4.

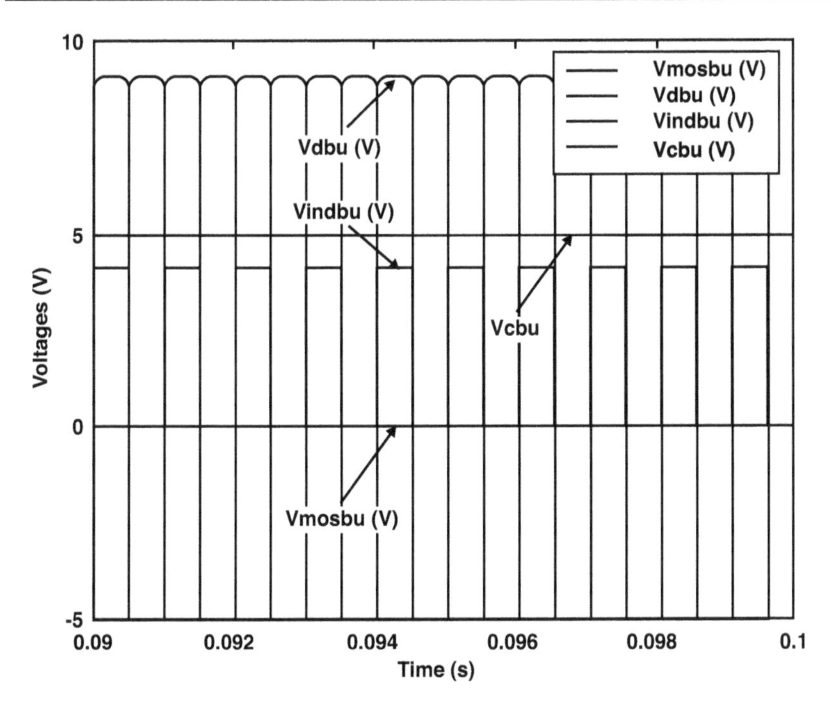

Figure 5.14 Plots of different voltages as per parabu#44.

Specification # 3

$V_{sbu} = 9$ V; $V_{cbu} = 5$ V; $D_{bu} = 5/9$; $I_{Lbu} = 1$ A; $\Delta I_{indbu} = 300$ mA (30% of the full load current); $\Delta V_{cbu} = 300$ mV (6% of the output voltage).

Specification # 4

$V_{sbu} = 9$ V; $V_{cbu} = 5$ V; $D_{bu} = 5/9$; $I_{Lbu} = 1$ A; $\Delta I_{indbu} = 20$ mA (2 % of the full load current); $\Delta V_{cbu} = 30$ mV (0.06 % of the output voltage).

5.4 CRITICAL VALUES OF INDUCTANCE AND CAPACITANCE

To maintain continuous current in the inductor one may calculate the values of critical inductance and capacitance as follows:

Specification # Critical

Let,

$V_{sbu} = 9$ V; $V_{cbu} = 5$ V; $D_{bu} = 5/9$; $f_{bu} = 1$ kHz; $R_{bu} = 5\ \Omega$

For the above specification,
The critical value of the inductance,

$$L_{cbu} = \frac{(1 - D_{bu})R_{bu}}{2f_{bu}}$$

Substituting the values in Eqn. 5.12, one may get,

$$L_{cbu} = \frac{\left(1 - \frac{5}{9}\right) \times 5}{2 \times 1 \times 10^3} = 0.0011\text{H}$$

The critical value of the capacitor,

$$C_{cbu} = \frac{(1 - D_{bu})}{16 L_{cbu} f_{bu}^{\ 2}}$$

Substituting the values in Eqn. 5.13, one may get,

$$C_{cbu} = \frac{(1 - D_{bu})}{16 L_{cbu} f_{bu}^{\ 2}} = \frac{\frac{4}{9}}{16 \times 0.0011 \times \left(10^3\right)^2} = 25.2\mu\text{F}$$

The variations of the inductor current and the capacitor voltage for a frequency of 1 kHz with critical values of inductance and capacitance are plotted in Figure 5.15 and Figure 5.16, respectively; one may note that inductor

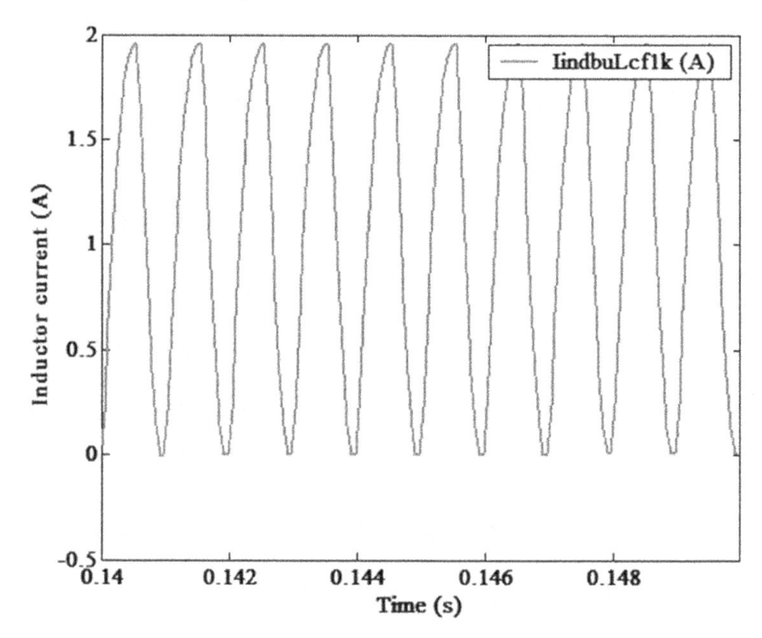

Figure 5.15 Plot of inductor current as per specification # critical.

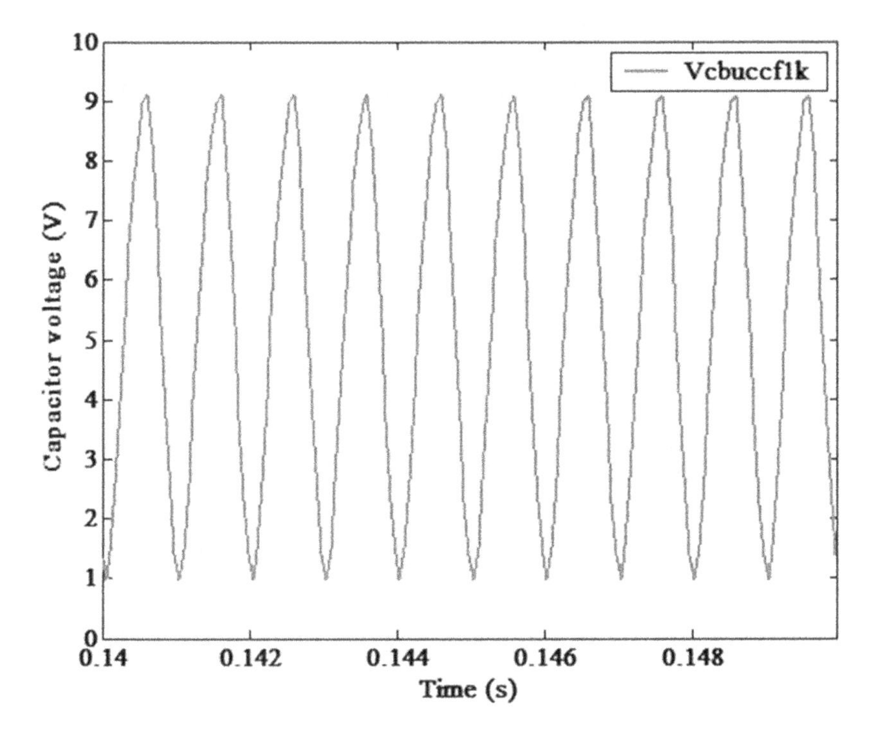

Figure 5.16 Plot of capacitor voltage as per specification # critical.

current switches from 0 A to 1.9 A and capacitor voltage switches from 1 V to 9 V.

BIBLIOGRAPHY

Bimbra P. S., Power electronics (4th ed.). India: Khanna Publishers, 2006.

Bose B. K., Modern power electronics and AC drives (1st ed.). India: Pearson Education Inc., 2002.

Bradley D. A., Power electronics (2nd ed.). England: Chapman & Hall, 1995.

Edminister J. A., Schaum's outline of theory and problems of electric circuits (Asian student edition). Singapore: McGraw-Hill International Book Company, 1983.

George M., Basu K. P., and Younis M. A. A., Digital simulation of static power converters using power system blockset (1st ed.). Germany: LAP LAMBERT Academic Publishing, 2012.

Hadi S., Power system analysis (2nd ed.). Singapore: McGraw-Hill Education (Asia), 2004.

Hambley A. R., Electrical engineering principles & applications (3rd ed., International). USA: Pearson Education International, 2005.

Hughes E., Electrical and electronics technology (10th ed.). UK: Pearson Education Limited, 2008.

MATLAB Education Seminar. Malaysia: International Islamic University, 2003.

Mehta V. K. and Mehta R., Principles of electrical engineering and electronics (multicolour illustrative edition, 2nd ed.). India: S. Chand & Company Ltd., 2006.
Mohan N., Power electronics modeling simplified using PSPICE (Release 9). Canada: University of Minnesota, 2002.
Mohan N., Undeland T. M., and Robbins W. P., Power electronics: converters, applications and design (3rd ed.). USA: John Wiley & Sons Inc., 2003.
Part-Enander E., Sjoberg A., Melin B., and Ishaksson P., The MATLAB handbook (1st ed.). USA: Addison Wesley, 1996.
Petruzella F. D., Industrial electronics (1st ed.). Electrical & electronic technology series. Singapore: McGraw-Hill International Editions, 1996.
Rai, H. C., Industrial and power electronics (1st ed.). India: Umesh Publications, 1987.
Rashid M. H., 2 days course on power electronics and its applications. Malaysia: Universiti Putra Malaysia, 2004.
Rashid M. H., Power electronics circuits, devices and applications (3rd ed.). India: Pearson Education Inc., 2004.
Rashid M. H., Tutorial on design and analysis of power converters. Malaysia: Universiti Putra Malaysia, 2002.
Sen P. C., Principles of electric machines and power electronics (2nd ed.). USA: John Wiley & Sons Inc., 1999.
Subramanyam V., Power electronics (1st ed.). India: New Age International Private Ltd., 2003.
The MathWorks Inc., Power system blockset for use with Simulink. USA: The MathWorks Inc., 2000.
Theraja A. K. and Theraja B. L., A textbook of electrical technology (25th ed.). India: S. Chand & Company Ltd., 2008.
Umanand L., Switched mode power conversion, online http://nptel.ac.in/, 2014.
Wildi T., Electrical machines, drives, and power systems (4th ed.). USA: Prentice Hall International Inc., 2000.

Chapter 6

Mathematical modelling and stability analysis of a buck converter

6.1 INTRODUCTION

Mathematical models describing a system with sets of equations are used for design, analysis, and control. State-space models use first order differential equations to model any dynamic systems. One may use large signal average models, steady state average models, and small signal average models to analyze the stability of the system under different operating conditions.

6.2 STATE-SPACE MODEL OF A BUCK CONVERTER[1]

As the power circuit of the buck converter is different in each mode, one may obtain the state-space models in each mode separately and then the average model should be considered for analysis.

6.2.1 State-space model of a buck converter in mode 1

The circuit topology of a buck converter in mode 1 with the switch S_{1bu} in the ON position for the state-space approach is shown in Figure 6.1.

The voltage across the inductor in Figure 6.1 could be expressed as:

$$L_{bu}\frac{d}{dt}\left(i_{indbu}\right) = V_{sbu} - v_{cbu} \tag{6.1}$$

rewriting Eqn. 6.1 one may get,

$$\frac{d}{dt}\left(i_{indbu}\right) = -\frac{v_{cbu}}{L_{bu}} + \frac{V_{sbu}}{L_{bu}} \tag{6.2}$$

DOI: 10.1201/9781003511236-6

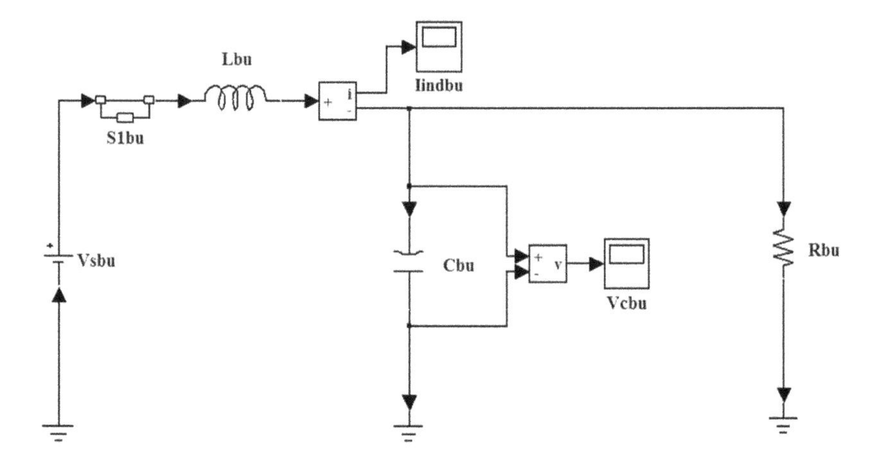

Figure 6.1 Circuit topology of a buck converter in mode 1 for state-space approach.

And the current through the inductor in Figure 6.1 could be expressed as:

$$i_{indbu} = C_{bu}\frac{d}{dt}(v_{cbu}) + \frac{v_{cbu}}{R_{bu}} \tag{6.3}$$

rewriting Eqn. 6.3, one may get,

$$\frac{d}{dt}(v_{cbu}) = \frac{i_{indbu}}{C_{bu}} - \frac{v_{cbu}}{R_{bu}C_{bu}} \tag{6.4}$$

Taking i_{indbu} and v_{cbu} as state variables and writing state equations in matrix form, one may get the state-space model of the buck converter in mode 1 as:

$$\begin{bmatrix} \frac{d}{dt}(i_{indbu}) \\ \frac{d}{dt}(v_{cbu}) \end{bmatrix} = \begin{bmatrix} 0 & \frac{-1}{L_{bu}} \\ \frac{1}{C_{bu}} & \frac{-1}{R_{bu}C_{bu}} \end{bmatrix} \begin{bmatrix} i_{indbu} \\ v_{cbu} \end{bmatrix} + \begin{bmatrix} \frac{1}{L_{bu}} \\ 0 \end{bmatrix} [v_{sbu}] \tag{6.5a}$$

$$[v_{cbu}] = [0 \quad 1]\begin{bmatrix} i_{indbu} \\ v_{cbu} \end{bmatrix} + [0][v_{sbu}] \tag{6.5b}$$

Comparing Eqn. (6.5a) with Eqn. (2.16), one may get,

$$A_{sbu1} = \begin{bmatrix} 0 & \frac{-1}{L_{bu}} \\ \frac{1}{C_{bu}} & \frac{-1}{R_{bu}C_{bu}} \end{bmatrix} \tag{6.6}$$

and

$$B_{sbu1} = \begin{bmatrix} \frac{1}{L_{bu}} \\ 0 \end{bmatrix} \tag{6.7}$$

Comparing Eqn. (6.5b) with Eqn. (2.17), one may get,

$$C_{sbu1} = \begin{bmatrix} 0 & 1 \end{bmatrix} \tag{6.8}$$

and

$$D_{sbu1} = [0] \tag{6.9}$$

6.2.2 State-space model of a buck converter in mode 2

The circuit topology of a buck converter in mode 2 with the switch S_{2bu} in the ON position for the state-space approach is shown in Figure 6.2.

The voltage across the capacitor in Figure 6.2 could be expressed as:

$$-L_{bu} \frac{d}{dt}(i_{indbu}) = v_{cbu} \tag{6.10}$$

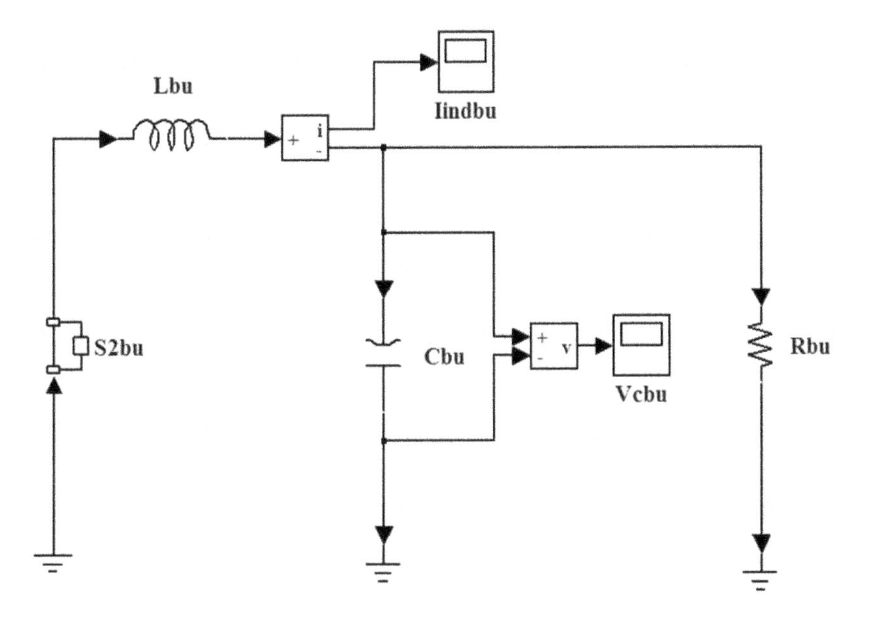

Figure 6.2 Circuit topology of a buck converter in mode 2 for state-space approach.

and current through the inductor in Figure 6.2 could be expressed as:

$$i_{indbu} = C_{bu}\frac{d}{dt}(v_{cbu}) + \frac{v_{cbu}}{R_{bu}} \quad (6.11)$$

Taking i_{indbu} and v_{cbu} as state variables and writing state equations in matrix form, one may get the state-space model of the buck converter in mode 2 as:

$$\begin{bmatrix} \frac{d}{dt}(i_{indbu}) \\ \frac{d}{dt}(v_{cbu}) \end{bmatrix} = \begin{bmatrix} 0 & \frac{-1}{L_{bu}} \\ \frac{1}{c_{bu}} & \frac{-1}{R_{bu}C_{bu}} \end{bmatrix} \begin{bmatrix} i_{indbu} \\ v_{cbu} \end{bmatrix} + \begin{bmatrix} 0 \\ 0 \end{bmatrix} [v_{sbu}] \quad (6.12a)$$

$$[v_{cbu}] = [0 \quad 1] \begin{bmatrix} i_{indbu} \\ v_{cbu} \end{bmatrix} + [0][v_{sbu}] \quad (6.12b)$$

Comparing Eqn. (6.12a) with Eqn. (2.16), one may get,

$$A_{sbu2} = \begin{bmatrix} 0 & \frac{-1}{L_{bu}} \\ \frac{1}{C_{bu}} & \frac{-1}{R_{bu}C_{bu}} \end{bmatrix} \quad (6.13)$$

and

$$B_{sbu2} = \begin{bmatrix} 0 \\ 0 \end{bmatrix} \quad (6.14)$$

Comparing Eqn. (6.12b) with Eqn. (2.17), one may get,

$$C_{sbu2} = [0 \quad 1] \quad (6.15)$$

and

$$D_{sbu2} = [0] \quad (6.16)$$

6.2.3 State-space average model of a buck converter

The duty cycle d_{bu} is the fraction of the time period during which switch S_{1bu} conducts and $1\text{-}d_{bu}$ is the fraction of the time period during which switch S_{1bu} is OFF.

From Eqn. 5.3, one may get,

$$1 - d_{bu} = 1 - \frac{T_{on}}{T_s} = \frac{T_{off}}{T_s} \quad (6.17)$$

So the average value of A_{sbu} is

$$A_{sbu} = \frac{A_{sbu1} \times d_{bu}T_s + A_{sbu2} \times (1 - d_{bu})T_s}{T_s} \tag{6.18}$$

From Eqn. 6.6 and Eqn. 6.13, one may get,

$$A_{sbu1} = A_{sbu2} \tag{6.19}$$

$$\therefore A_{sbu} = \begin{bmatrix} 0 & \dfrac{-1}{L_{bu}} \\ \dfrac{1}{C_{bu}} & \dfrac{-1}{R_{bu}C_{bu}} \end{bmatrix} \tag{6.20}$$

Likewise the average value of B_{sbu} is

$$B_{sbu} = \frac{B_{sbu1} \times d_{bu}T_s + B_{sbu2} \times (1 - d_{bu})T_s}{T_s} \tag{6.21}$$

From Eqn. 6.7, Eqn. 6.14 and Eqn. 6.21, one may get,

$$B_{sbu} = \begin{bmatrix} \dfrac{d_{bu}}{L_{bu}} \\ 0 \end{bmatrix} + \begin{bmatrix} 0 \\ 0 \end{bmatrix} = \begin{bmatrix} \dfrac{d_{bu}}{L_{bu}} \\ 0 \end{bmatrix} \tag{6.22}$$

Similarly,

$$C_{sbu} = [0 \;\; 1] \tag{6.23}$$

and

$$D_{sbu} = [0] \tag{6.24}$$

Thus, the state-space average model of the buck converter could be written as follows:

$$\begin{bmatrix} \dfrac{d}{dt}(i_{indbu}) \\ \dfrac{d}{dt}(v_{cbu}) \end{bmatrix} = \begin{bmatrix} 0 & \dfrac{-1}{L_{bu}} \\ \dfrac{1}{C_{bu}} & \dfrac{-1}{R_{bu}C_{bu}} \end{bmatrix} \begin{bmatrix} i_{indbu} \\ v_{cbu} \end{bmatrix} + \begin{bmatrix} \dfrac{d_{bu}}{L_{bu}} \\ 0 \end{bmatrix} [v_{sbu}] \tag{6.25a}$$

$$[v_{cbu}] = [0 \;\; 1] \begin{bmatrix} i_{indbu} \\ v_{cbu} \end{bmatrix} + [0][v_{sbu}] \tag{6.25b}$$

6.2.3.1 Large signal average model of a buck converter

Under transient conditions,

$$d_{bu} = D_{bu} + \hat{d}_{bu} \tag{6.26}$$

where D_{bu} is the steady state value of the duty cycle $\hat{d}_{bu}$ is the deviation in the neighbourhood of the operating point.

Considering the variations in the neighbourhood of the operating point, substituting Eqn. 6.26 in Eqn. 6.18, one may get the average large signal model of the buck converter as follows:

$$A_{sbulr} = \frac{A_{sbu1} \times \left(D_{bu} + \hat{d}_{bu}\right)T_s + A_{sbu2} \times \left(1 - \left(D_{bu} + \hat{d}_{bu}\right)\right)T_s}{T_s} \tag{6.27}$$

However since

$$A_{sbu1} = A_{sbu2}$$

$$A_{sbulr} = \begin{bmatrix} 0 & \dfrac{-1}{L_{bu}} \\ \dfrac{1}{C_{bu}} & \dfrac{-1}{R_{bu}C_{bu}} \end{bmatrix} \tag{6.28}$$

Likewise,

$$B_{sbulr} = \frac{B_{sbu1} \times \left(D_{bu} + \hat{d}_{bu}\right)T_s + Bs_{bu2} \times \left(1 - \left(D_{bu} + \hat{d}_{bu}\right)\right)T_s}{T_s} \tag{6.29}$$

Substituting Eqn. 6.7 and Eqn. 6.14 in Eqn. 6.29, one may get,

$$B_{sbulr} = \begin{bmatrix} \dfrac{D_{bu} + \hat{d}_{bu}}{L_{bu}} \\ 0 \end{bmatrix} \tag{6.30}$$

Similarly,

$$C_{sbulr} = \begin{bmatrix} 0 & 1 \end{bmatrix} \tag{6.31}$$

and

$$D_{sbulr} = [0] \tag{6.32}$$

Considering the effect of large variations in the input voltage and the duty cycle, one may write the dynamic equation as follows:

$$\begin{bmatrix} \frac{d}{dt}\left(I_{indbu}+\hat{i}_{indbu}\right) \\ \frac{d}{dt}\left(V_{cbu}+\hat{v}_{cbu}\right) \end{bmatrix} = \begin{bmatrix} 0 & \frac{-1}{L_{bu}} \\ \frac{1}{C_{bu}} & \frac{-1}{R_{bu}C_{bu}} \end{bmatrix} \begin{bmatrix} I_{indbu}+\hat{i}_{indbu} \\ V_{cbu}+\hat{v}_{cbu} \end{bmatrix} + \begin{bmatrix} \frac{D_{bu}+\hat{d}_{bu}}{L_{bu}} \\ 0 \end{bmatrix} \left[V_{sbu}+\hat{v}_{sbu}\right] \quad (6.33)$$

Rearranging Eqn. 6.33 one may get,

$$\begin{bmatrix} \frac{d}{dt}\left(I_{indbu}+\hat{i}_{indbu}\right) \\ \frac{d}{dt}\left(V_{cbu}+\hat{v}_{cbu}\right) \end{bmatrix}$$

$$= \begin{bmatrix} 0 & \frac{-1}{L_{bu}} \\ \frac{1}{C_{bu}} & \frac{-1}{R_{bu}C_{bu}} \end{bmatrix} \begin{bmatrix} I_{indbu}+\hat{i}_{indbu} \\ V_{cbu}+\hat{V}_{cbu} \end{bmatrix} + \begin{bmatrix} \frac{D_{bu}}{L_{bu}} \\ 0 \end{bmatrix} \left[V_{sbu}\right] + \begin{bmatrix} \frac{\hat{d}_{bu}}{L_{bu}} \\ 0 \end{bmatrix} \left[V_{sbu}\right] \quad (6.34)$$

$$+ \begin{bmatrix} \frac{D_{bu}}{L_{bu}} \\ 0 \end{bmatrix} \left[\hat{v}_{sbu}\right] + \begin{bmatrix} \frac{\hat{d}_{bu}}{L_{bu}} \\ 0 \end{bmatrix} \left[\hat{v}_{sbu}\right]$$

Neglecting the multiplication of the two small signals $\hat{d}_{bu}$ and $\hat{v}_{sbu}$, the large signal average model could be written as:

$$\begin{bmatrix} \frac{d}{dt}\left(I_{indbu}+\hat{i}_{indbu}\right) \\ \frac{d}{dt}\left(V_{cbu}+\hat{v}_{cbu}\right) \end{bmatrix}$$

$$= \begin{bmatrix} 0 & \frac{-1}{L_{bu}} \\ \frac{1}{C_{bu}} & \frac{-1}{R_{bu}C_{bu}} \end{bmatrix} \begin{bmatrix} I_{indbu}+\hat{i}_{indbu} \\ V_{cbu}+\hat{V}_{cbu} \end{bmatrix} + \begin{bmatrix} \frac{D_{bu}}{L_{bu}} \\ 0 \end{bmatrix} \left[V_{sbu}\right] + \begin{bmatrix} \frac{D_{bu}}{L_{bu}} & \frac{V_{sbu}}{L_{bu}} \\ 0 & 0 \end{bmatrix} \begin{bmatrix} \hat{v}_{sbu} \\ \hat{d}_{bu} \end{bmatrix} \quad (6.35a)$$

$$\left[V_{cbu}+\hat{V}_{cbu}\right] = \left[0 \;\; 1\right] \begin{bmatrix} I_{indbu}+\hat{i}_{indbu} \\ V_{cbu}+\hat{V}_{cbu} \end{bmatrix} \quad (6.35b)$$

6.2.3.2 Steady state average model of a buck converter

Under steady state condition,

$$d_{bu} = D_{bu} \tag{6.36}$$

One may also note that under steady state condition,

$$\frac{dx}{dt} = A_s x + B_s u = 0 \tag{6.37}$$

hence, under steady state condition,

$$\begin{bmatrix} \frac{d}{dt}(I_{indbu}) \\ \frac{d}{dt}(V_{cbu}) \end{bmatrix} = \begin{bmatrix} 0 & \frac{-1}{L_{bu}} \\ \frac{1}{C_{bu}} & \frac{-1}{R_{bu}C_{bu}} \end{bmatrix} \begin{bmatrix} I_{indbu} \\ V_{cbu} \end{bmatrix} + \begin{bmatrix} \frac{D_{bu}}{L_{bu}} \\ 0 \end{bmatrix} [V_{sbu}] = 0 \tag{6.38}$$

$$[V_{cbu}] = [0 \quad 1] \begin{bmatrix} I_{indbu} \\ V_{cbu} \end{bmatrix} \tag{6.39}$$

From Eqn. 6.38, one may get,

$$A_{sbust} = \begin{bmatrix} 0 & \frac{-1}{L_{bu}} \\ \frac{1}{C_{bu}} & \frac{-1}{R_{bu}C_{bu}} \end{bmatrix} \tag{6.40}$$

and

$$B_{sbust} = \begin{bmatrix} \frac{D_{bu}}{L_{bu}} \\ 0 \end{bmatrix} \tag{6.41}$$

From Eqn. 6.39, one may get,

$$C_{sbust} = [0 \quad 1] \tag{6.42}$$

and

$$D_{sbust} = [0] \tag{6.43}$$

Eqn. 6.37 implies that,

$$A_s x + B_s u = 0 \tag{6.44}$$

$$\therefore x = A_s^{-1}(-B_s u) = -A_s^{-1} B_s u \tag{6.45}$$

$$\therefore y = C_s x + D_s u = C_s\left(-A_s^{-1} B_s\right) + D_s u \tag{6.46}$$

Substituting x, u, A_{sbust} and B_{sbust} in Eqn. 6.45, one may get,

$$\begin{bmatrix} I_{indbu} \\ V_{cbu} \end{bmatrix} = -\begin{bmatrix} 0 & \frac{-1}{L_{bu}} \\ \frac{1}{C_{bu}} & \frac{-1}{R_{bu} C_{bu}} \end{bmatrix}^{-1} \begin{bmatrix} \frac{D_{bu}}{L_{bu}} \\ 0 \end{bmatrix} [V_{sbu}] \tag{6.47}$$

Substituting x, u, y, C_{sbust} and D_{sbust} in Eqn. 6.46, one may get,

$$\therefore [V_{cbu}] = [0 \quad 1] \begin{bmatrix} I_{indbu} \\ V_{cbu} \end{bmatrix} + [0][V_{sbu}] \tag{6.48}$$

Substituting the expression of

$$\begin{bmatrix} I_{indbu} \\ V_{cbu} \end{bmatrix}$$

given by Eqn. 6.47 in Eqn. 6.48, one may get,

$$[V_{cbu}] = -[0 \quad 1] \begin{bmatrix} 0 & \frac{-1}{L_{bu}} \\ \frac{1}{C_{bu}} & \frac{-1}{R_{bu} C_{bu}} \end{bmatrix}^{-1} \begin{bmatrix} \frac{D_{bu}}{L_{bu}} \\ 0 \end{bmatrix} [V_{sbu}] \tag{6.49}$$

6.2.3.3 Small signal average model of a buck converter

To get the dynamic equation of the small signal model one may apply the steady state condition $\frac{dx}{dt} = A_s x + B_s u = 0$ in the large signal model given by Eqn. 6.35a. One may get,

$$\begin{bmatrix} \frac{d}{dt}\left(\hat{i}_{indbu}\right) \\ \frac{d}{dt}\left(\hat{v}_{cbu}\right) \end{bmatrix} = \begin{bmatrix} 0 & \frac{-1}{L_{bu}} \\ \frac{1}{c_{bu}} & \frac{-1}{R_{bu} c_{bu}} \end{bmatrix} \begin{bmatrix} \hat{i}_{indbu} \\ \hat{v}_{cbu} \end{bmatrix} + \begin{bmatrix} \frac{D_{bu}}{L_{bu}} & \frac{V_{sbu}}{L_{bu}} \\ 0 & 0 \end{bmatrix} \begin{bmatrix} \hat{v}_{sbu} \\ \hat{d}_{bu} \end{bmatrix} \tag{6.50}$$

Neglecting the variation in the input voltage in Eqn. 6.50, one may get,

$$\begin{bmatrix} \frac{d}{dt}\left(\hat{i}_{indbu}\right) \\ \frac{d}{dt}\left(\hat{v}_{cbu}\right) \end{bmatrix} = \begin{bmatrix} 0 & \frac{-1}{L_{bu}} \\ \frac{1}{c_{bu}} & \frac{-1}{R_{bu}C_{bu}} \end{bmatrix} \begin{bmatrix} \hat{i}_{indbu} \\ \hat{v}_{cbu} \end{bmatrix} + \begin{bmatrix} \frac{V_{sbu}}{L_{bu}} \\ 0 \end{bmatrix} \begin{bmatrix} \hat{d}_{bu} \end{bmatrix} \tag{6.51}$$

From Eqn. 6.35b, the small signal variation in the output voltage,

$$\begin{bmatrix} \hat{v}_{cbu} \end{bmatrix} = [01] \begin{bmatrix} \hat{i}_{indbu} \\ \hat{v}_{cbu} \end{bmatrix} \tag{6.52}$$

Thus,

$$A_{sbusm} = \begin{bmatrix} 0 & \frac{-1}{L_{bu}} \\ \frac{1}{C_{bu}} & \frac{-1}{R_{bu}C_{bu}} \end{bmatrix} \tag{6.53}$$

$$B_{sbusm} = \begin{bmatrix} \frac{V_{sbu}}{L_{bu}} \\ 0 \end{bmatrix} \tag{6.54}$$

$$C_{sbusm} = \begin{bmatrix} 0 & 1 \end{bmatrix} \tag{6.55}$$

And

$$D_{sbusm} = [0] \tag{6.56}$$

6.3 STABILITY ANALYSIS OF A BUCK CONVERTER

Stability of a plant may change with variations in the operating conditions. Consider a negative feedback closed loop control system as shown in Figure 6.3 with transfer functions expressed as functions of numerator and denominator polynomials; let

G_p (s)—transfer function of the plant (n_p (s)/d_p (s))
G_c (s)—transfer function of the controller (n_c (s)/d_c (s))
G_h (s)—transfer function of the feedback path (n_h (s)/d_h (s))

One may note that for the closed loop system to be stable, the loop transfer function (G_p (s) G_c (s) G_h (s)) should not have any zeros or poles on the right half of the s-plane and both Gm and Pm should be positive.

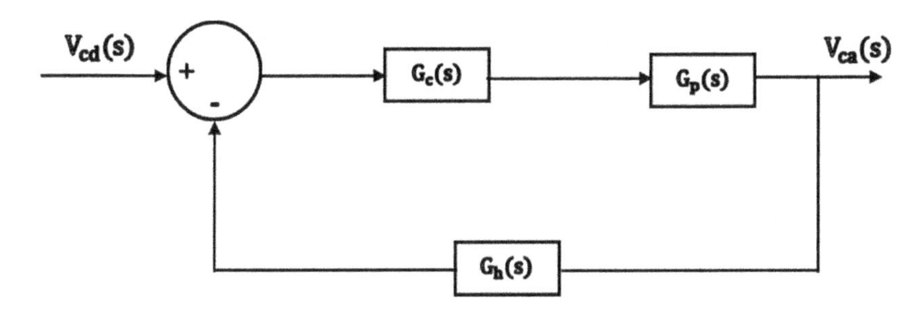

Figure 6.3 Negative feedback closed loop control system.

Both large signal and small signal average models are taken into consideration to analyze the stability of the buck converter; variation in the input voltage is considered to obtain the transfer function of the large signal average model and variation in the duty cycle is considered to obtain the transfer function of the small signal average model.

6.3.1 Stability analysis of the large signal average model

Consider a buck converter circuit with the following specifications:

V_{sbu} = 9 V; D_{bu} = 5/9; L_{bu} = 10 mH; C_{bu} = 100 μF; R_{bu} = 5 Ω; f_{bu} = 1 kHz.

One may get the transfer function model of the large signal average model of the buck converter as follows:

Substituting the values in Eqn. (6.25a) one may get,

$$\begin{bmatrix} \frac{d}{dt}\left(i_{indbu}\right) \\ \frac{d}{dt}\left(v_{cbu}\right) \end{bmatrix} = \begin{bmatrix} 0 & -100 \\ 10000 & -2000 \end{bmatrix}\begin{bmatrix} i_{indbu} \\ v_{cbu} \end{bmatrix} + \begin{bmatrix} 55.556 \\ 0 \end{bmatrix}\left[v_{sbu}\right] \tag{6.57}$$

From Eqn. 6.57 one may get,

$$\frac{d}{dt}\left(i_{indbu}\right) = -100v_{cbu} + 55.556v_{sbu} \tag{6.58}$$

or in otherwards,

$$\frac{d}{dt}\left(i_{indbu}(t)\right) = -100v_{cbu}(t) + 55.556v_{sbu}(t) \tag{6.59}$$

Taking the Laplace transform of Eqn. 6.59, one may get,

$$sI_{indbu}(s) = -100V_{cbu}(s) + 55.556V_{sbu}(s) \tag{6.60}$$

From Eqn. 6.57 one may get,

$$\frac{d}{dt}(v_{cbu}) = 10000 i_{indbu} - 2000 v_{cbu} \tag{6.61}$$

or in otherwards,

$$\frac{d}{dt}(v_{cbu}(t)) = 10000 i_{indbu}(t) - 2000 v_{cbu}(t) \tag{6.62}$$

Taking the Laplace transform of Eqn. 6.62, one may get,

$$sV_{cbu}(s) = 10000 I_{indbu}(s) - 2000 V_{cbu}(s) \tag{6.63}$$

From Eqn. 6.60 one may get,

$$I_{indbu}(s) = \frac{[-100V_{cbu}(s) + 55.556V_{sbu}(s)]}{s} \tag{6.64}$$

Substituting the expression of $I_{indbu}(s)$ from Eqn. 6.64 in Eqn. 6.63, one may get,

$$sV_{cbu}(s) = 10000 \frac{[-100V_{cbu}(s) + 55.556V_{sbu}(s)]}{s} - 2000V_{cbu}(s) \tag{6.65}$$

Rearranging Eqn. 6.65, one may get,

$$s^2 V_{cbu}(s) = 10000[-100V_{cbu}(s) + 55.556V_{sbu}(s)] - s2000V_{cbu}(s) \tag{6.66}$$

Rearranging Eqn. 6.66, one may get,

$$s^2 V_{cbu}(s) + 10^6 V_{cbu}(s) + s2000V_{cbu}(s) = 55.556 \times 10^4 V_{sbu}(s) \tag{6.67}$$

From Eqn. 6.67, one may get,

$$\frac{V_{cbu}(s)}{V_{sbu}(s)} = \frac{55.556 \times 10^4}{s^2 + 2000s + 10^6} = \frac{55.556 \times 10^4}{(s + 1000)^2} \tag{6.68}$$

Analysis of Eqn. 6.68 shows that for the above system there are no zeros and the poles are located at s = –1000.

From Eqn. 6.68, one may get,

$$V_{cbu}(s) = \frac{55.556 \times 10^4}{(s + 1000)^2} \times V_{sbu}(s) \tag{6.69}$$

mcode6–1

MATLAB® code ss2tf used in mcode6–1 obtains the transfer function model from the state-space model; coding without correct format in mcode6–2 would result a wrong transfer function model as compared to the correct one.

mcode6–1

```
>> Vsbu = 9; Dbu = 5/9; Lbu = 10e-3; Cbu = 100e–6; Rbu = 5; Pdo = 5;
>> Asbulr = [0–1/Lbu;1/Cbu –1/(Rbu*Cbu)];
>> Bsbulr = [Dbu/Lbu;0];
>> Csbulr = [0 1]; Dsbulr = [0];
>> format long G
>> [nplr,dplr] = ss2tf(Asbulr,Bsbulr,Csbulr,Dsbulr);
>> nplr = round(nplr);
>> dplr = round(dplr);
>> tf(nplr,dplr)
Transfer function:
555556
-------------------
s^2 + 2000 s + 1e006
mcode6–2
>> Vsbu = 9; Dbu = 5/9; Lbu = 10e-3; Cbu = 100e-6; Rbu = 5; Pdo = 5;
>> Asbulr = [0–1/Lbu;1/Cbu –1/(Rbu*Cbu)];
>> Bsbulr = [Dbu/Lbu;0];
>> Csbulr = [0 1]; Dsbulr = [0];
>> [nplr,dplr]=ss2tf(Asbulr,Bsbulr,Csbulr,Dsbulr)
nplr =
1.0e + 005 *
0 0.0000 5.5556
dplr =
1.0e + 005 *
0.0000 0.0200 10.0000
>> tf(nplr,dplr)
Transfer function:
4.547e–013 s + 5.556e005
-----------------------
s^2 + 2000 s + 1e006
```

Algorithm for stability analysis of the large signal average model

- Enter the values of input voltage (Vsbu), Duty cycle (Dsbu), Inductance (Lbu), Capacitance (Cbu), Load resistance (Rbu).
- Calculate the parameter matrix of the large signal average model

$$[A_{Sbulr}] = \begin{bmatrix} 0 & \frac{-1}{L_{bu}} \\ \frac{1}{C_{bu}} & \frac{-1}{R_{bu}C_{bu}} \end{bmatrix}$$

- Calculate the input matrix of the large signal average model

$$[B_{sbulr}] = \begin{bmatrix} \frac{D_{bu}}{L_{bu}} \\ 0 \end{bmatrix}$$

- Calculate the output matrix of the large signal model

$$[C_{sbulr}] = [0 \quad 1]$$

- Calculate the feed forward matrix of the large signal average model

$$[D_{sbulr}] = [0]$$

- Determine the transfer function model $\left(\frac{n_{plr}(s)}{d_{plr}(s)}\right)$ from the state-space model

$$(A_{sbulr}, B_{sbulr}, C_{sbulr}, D_{sbulr})$$

- Initialize controller gain to unity $\left(\frac{n_c(s)}{d_c(s)}\right) = \frac{1}{1}$
- Initialize feedback path transfer function to unity $\left(\frac{n_h(s)}{d_h(s)}\right) = \frac{1}{1}$
- Multiply the polynomials of large signal average model, controller and feedback path to get the loop transfer function
- Plot the pole-zero map
- If no poles or zeros on the right half of the s-plane then
    - o Plot the Bode diagram and obtain Gm and Pm else
    - o Obtain Gm and Pm
    - o Include the gain factor to achieve a stable system
    - o Plot the Bode diagram
- Plot the closed loop step response

The pole-zero map and Bode diagram of the loop transfer function of the large signal average model are shown in Figure 6.4 and Figure 6.5 respectively. As the system is stable with no zeros and poles on the right half of the s-plane and with both gain margin and phase margin infinity one may get a stable closed loop step response of the large signal average model as shown in Figure 6.6.

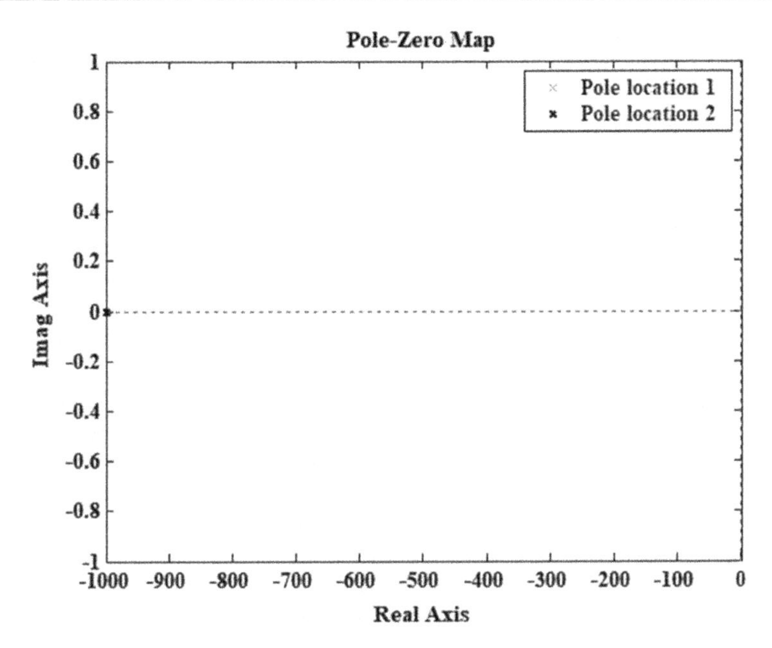

Figure 6.4 Pole-zero map of the loop transfer function of the large signal average model.

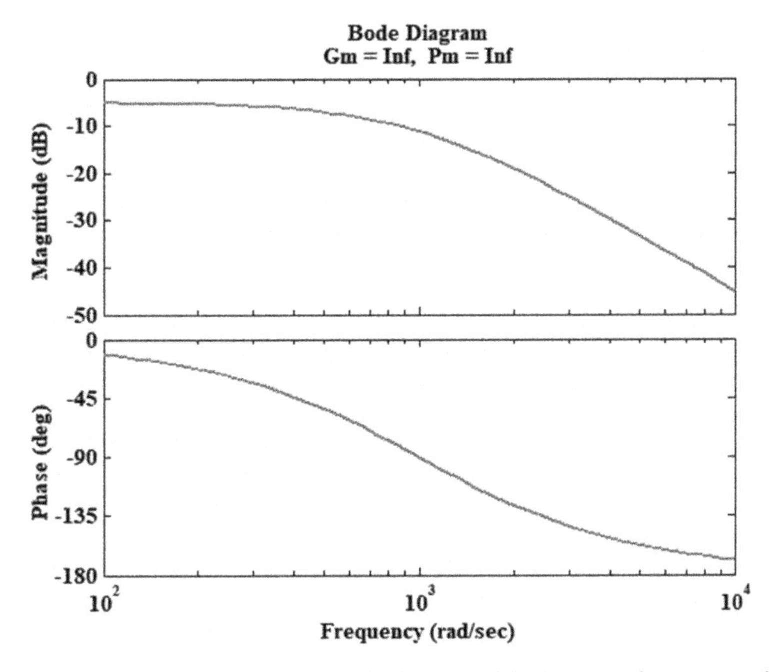

Figure 6.5 Bode diagram of the loop transfer function of the large signal average model.

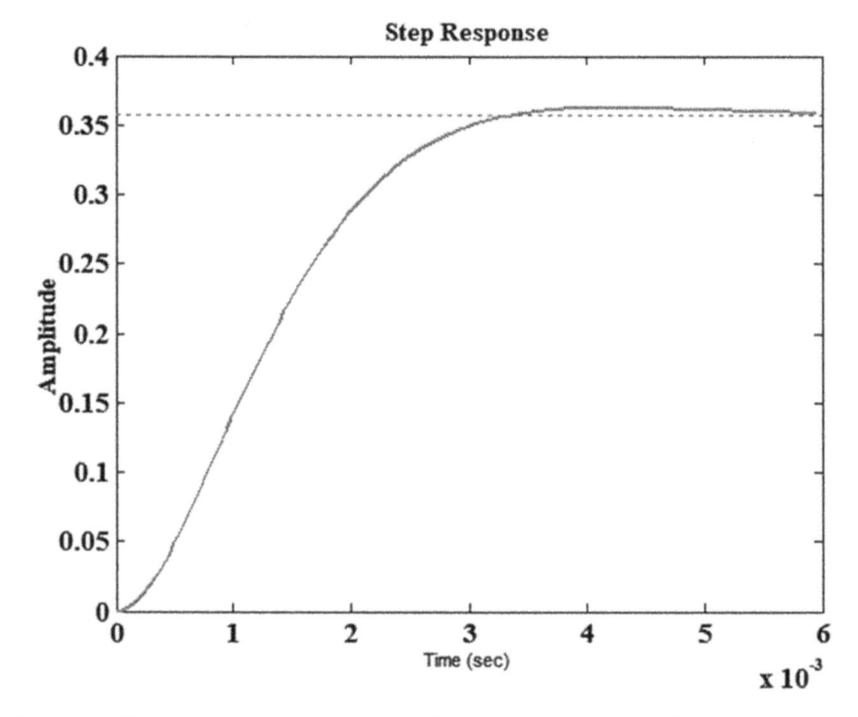

Figure 6.6 Closed loop step response of the large signal average model.

6.3.2 Stability analysis of the small signal average model

Consider a buck converter circuit having the following specifications:

V_{sbu} = 9 V; D_{bu} = 5/9; L_{bu} = 10 mH; C_{bu} = 100 μF; R_{bu} = 5 Ω; f_{bu} = 1 kHz.

Substituting the values in Eqn. (6.51), one may get,

$$\begin{bmatrix} \frac{d}{dt}\left(\hat{i}_{indbu}\right) \\ \frac{d}{dt}\left(\hat{v}_{cbu}\right) \end{bmatrix} = \begin{bmatrix} 0 & -100 \\ 10000 & -2000 \end{bmatrix} \begin{bmatrix} \hat{i}_{indbu} \\ \hat{v}_{cbu} \end{bmatrix} + \begin{bmatrix} 900 \\ 0 \end{bmatrix} \begin{bmatrix} \hat{d}_{bu} \end{bmatrix} \tag{6.70}$$

From Eqn. 6.70 one may get,

$$\frac{d}{dt}\left(\hat{i}_{indbu}\right) = -100\,\hat{v}_{cbu} + 900\,\hat{d}_{bu} \tag{6.71}$$

or in other words,

$$\frac{d}{dt}\left(\hat{i}_{indbu}(t)\right) = -100\,\hat{v}_{cbu}(t) + 900\,\hat{d}_{bu}(t) \tag{6.72}$$

Taking Laplace transform of Eqn. 6.72 one may get,

$$s\,\hat{I}_{indbu}(s) = -100\,\hat{V}_{cbu}(s) + 900\,\hat{D}_{bu}(s) \tag{6.73}$$

From Eqn. 6.70 one may get,

$$\frac{d}{dt}\left(\hat{v}_{cbu}\right) = 10000\,\hat{i}_{indbu} - 2000\,\hat{v}_{cbu} \tag{6.74}$$

or in other words,

$$\frac{d}{dt}\left(v_{cbu}(t)\right) = 10000\,\hat{i}_{indbu}(t) - 2000\,\hat{v}_{cbu}(t) \tag{6.75}$$

Taking Laplace transform of Eqn. 6.75, one may get,

$$s\,\hat{V}_{cbu}(s) = 10000\,\hat{I}_{indbu}(s) - 2000\,\hat{V}_{cbu}(s) \tag{6.76}$$

From Eqn. 6.76, one may get,

$$\hat{I}_{indbu}(s) = \frac{s\,\hat{V}_{cbu}(s) + 2000\,\hat{V}_{cbu}(s)}{10000} \tag{6.77}$$

Substituting the expression of $\hat{I}_{indbu}(s)$ from Eqn. 6.77 in Eqn. 6.73, one may get,

$$s\left(\frac{s\,\hat{V}_{cbu}(s) + 2000\,\hat{V}_{cbu}(s)}{10000}\right) = -100\,\hat{V}_{cbu}(s) + 900\,\hat{D}_{bu}(s) \tag{6.78}$$

Rearranging Eqn. 6.78, one may get,

$$\frac{\hat{V}_{cbu}(s)}{\hat{D}_{bu}(s)} = \frac{9\times10^6}{s^2 + 2000s + 10^6} = \frac{9\times10^6}{(s+1000)^2} \tag{6.79}$$

The pole-zero map and Bode plot shown in Figure 6.7 and Figure 6.8, respectively, illustrate that the closed loop response of the small signal average model is stable as depicted in Figure 6.9.

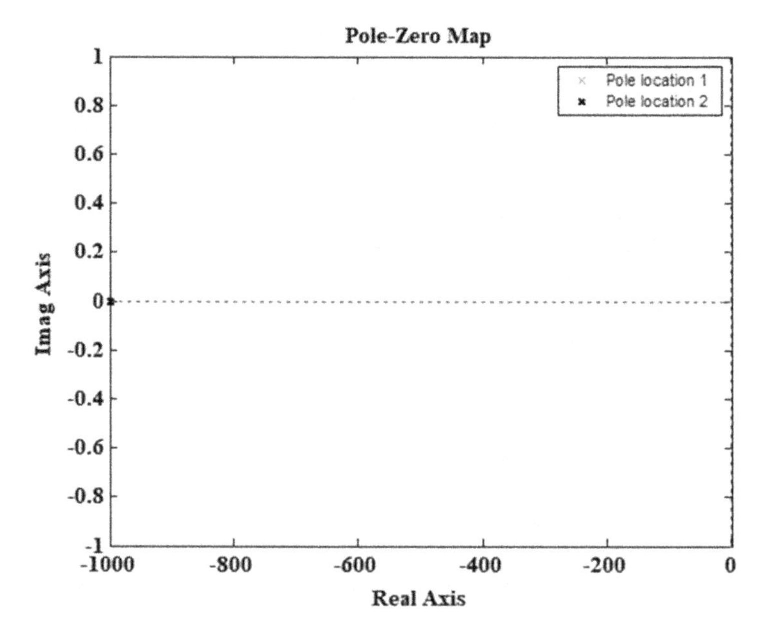

Figure 6.7 Pole-zero map of the loop transfer function of the small signal average model.

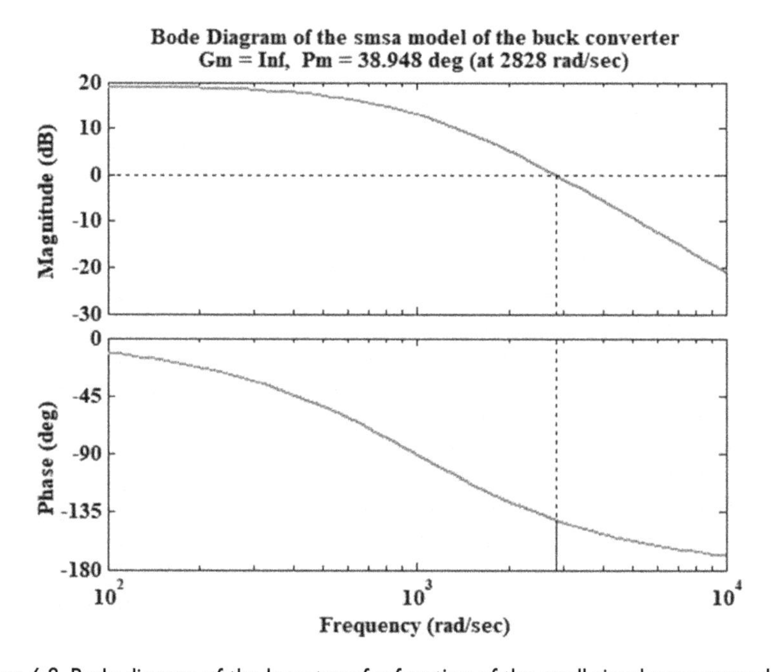

Figure 6.8 Bode diagram of the loop transfer function of the small signal average model.

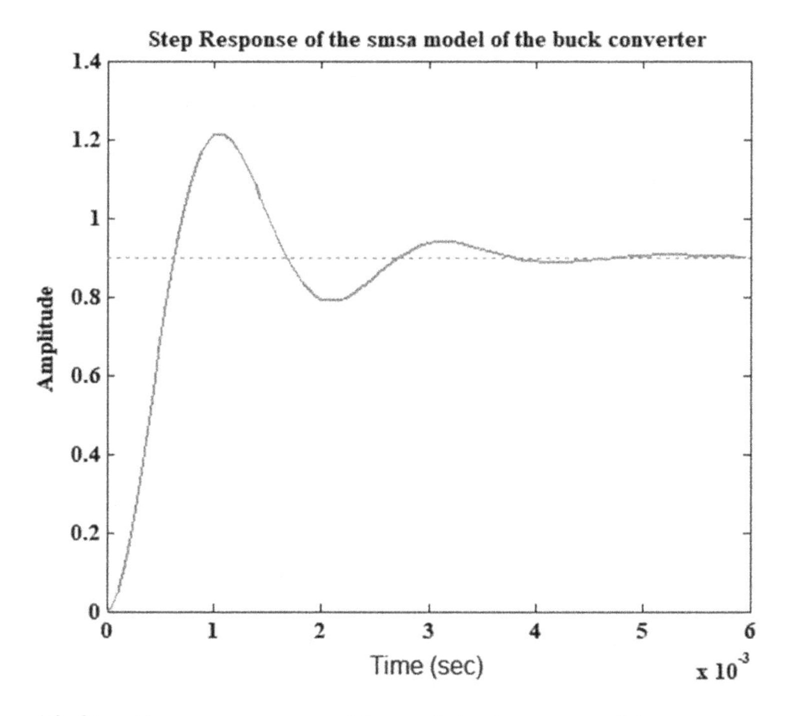

Figure 6.9 Closed loop step response of the small signal average model.

NOTE

1 Umanand L., Switched mode power conversion (nptel 2014).

BIBLIOGRAPHY

Benjamin C. K., Automatic control systems (5th ed.). India: Prentice Hall of India Private Ltd., 1989.

Dorf R. C. and Bishop R. H., Modern control systems (1st ed.). World student series. USA: Addison-Wesley, 1998.

Electrical systems/basic alternating (AC) theory/transfer function analysis, online http://control.com/textbook/ac-electricity/transfer-function-analysis/, 2022.

Hadi S., Power system analysis (2nd ed.). Singapore: McGraw-Hill Education (Asia), 2004.

Manke B. S., Linear control systems with MATLAB applications (8th ed.). India: Khanna Publishers, 2005.

Nagoor K. A., Control systems (1st ed.). India: R B A Publications, 1998.

Ogata K., Modern control engineering (3rd ed., International). USA: Prentice Hall International Inc., 1997.

Rashid M. H., Tutorial on design and analysis of power converters. Malaysia: Universiti Putra Malaysia, 2002.

Umanand L., Switched mode power conversion, online http://nptel.ac.in/, 2014.

Chapter 7

Performance analysis of the mathematical models of a buck converter

7.1 INTRODUCTION

One may analyze the performance of the mathematical models of the open loop and closed loop buck converters to verify the accuracy of the model. Partial fraction techniques could be used to obtain the time responses of the open loop buck converter using large signal average model and small signal average model, and a state-space approach could be used for the steady state average model. One may use an integral controller for the closed loop control of large signal average model to achieve the desired output performance. An algorithm is included to decide the gain of the controller using the root locus technique. Proportional controllers could be used for the control of small signal average models. One may use a Bode plot to determine the limiting value of the gain of the proportional controller; the algorithm included could be used to achieve the desired output performance.

7.2 PERFORMANCE ANALYSIS OF THE MATHEMATICAL MODELS OF THE OPEN LOOP BUCK CONVERTER

The effectiveness of the large signal average model, steady state average model, and small signal average model is discussed in this section. Transfer function models of the large signal average model and small signal average model are used to analyze the time response; initial value and final value theorems are applied for the analysis. Performance of the mathematical model has been verified using MATLAB® codes/SIMULINK models. An algorithm is included to analyze the performance of steady state average model using state-space approach.

DOI: 10.1201/9781003511236-7

7.2.1 Performance analysis of the large signal average model using transfer function

The output response of the large signal average model of the buck converter with variations in the input voltage could be derived using transfer function model as follows:

For a unit step input Eqn. (6.69) could be written as:

$$V_{cbu}(s) = \frac{55.556 \times 10^4}{(s+1000)^2} \times \frac{1}{s} \tag{7.1}$$

Using partial fractions Eqn. 7.1 could be written as:

$$\frac{55.556 \times 10^4}{s(s+1000)^2} = \frac{A_1}{s} + \frac{A_2}{(s+1000)} + \frac{A_3}{(s+1000)^2} \tag{7.2}$$

or in otherwards,

$$\frac{55.556 \times 10^4}{s(s+1000)^2} = \frac{A_1(s+1000)^2 + A_2 s(s+1000) + A_3 s}{s(s+1000)^2} \tag{7.3}$$

From Eqn. 7.3, one may get,

$$A_1\left(s^2 + 2000s + 10^6\right) + A_2 s(s+1000) + A_3 S = 55.556 \times 10^4 \tag{7.4}$$

From Eqn. 7.4, one may get,

$$s^2(A_1 + A_2) + s(2000A_1 + 1000A_2 + A_3) + 10^6 A_1 = 55.556 \times 10^4 \tag{7.5}$$

From Eqn. 7.5, one may get,

$$A_1 = 0.5556 \tag{7.6}$$

From Eqn. 7.5 and Eqn. 7.6, one may get,

$$A_1 + A_2 = 0 \quad \therefore A_2 = -A_1 = -0.5556 \tag{7.7}$$

From Eqn. 7.5, Eqn. 7.6 and Eqn. 7.7, one may get,

$$2000A_1 + 1000A_2 + A_3 = 2000 \times 0.5556 + 1000 \times -0.5556 + A_3 = 0 \tag{7.8}$$

From Eqn. 7.8, one may get,

$$A_3 = -555.6 \tag{7.9}$$

Substituting the values of A_1, A_2, A_3 in Eqn. 7.2, one may get,

$$\therefore V_{cbu}(s) = \frac{55.556 \times 10^4}{s(s+1000)^2} = \frac{0.5556}{s} - \frac{0.5556}{(s+1000)} - \frac{555.6}{(s+1000)^2} \tag{7.10}$$

Taking the inverse Laplace form of Eqn. 7.10, one may get,

$$\therefore v_{cbu}(t) = L^{-1}\left(\frac{0.5556}{s} - \frac{0.5556}{(s+1000)} - \frac{555.6}{(s+1000)^2}\right)$$

$$= 0.5556 - 0.5556e^{-1000t} - 555.6te^{-1000t} \tag{7.11}$$

When t = 0

$$v_{cbu}(t) = 0.5556 - 0.5556e^{-1000t} - 555.6t^{-1000t} = 0.5556 - 0.5556 = 0$$

One may also apply initial value theorem and final value theorem to calculate the initial value and final value of the capacitor voltage.

7.2.1.1 *Initial Value Theorem*

One may also apply initial value theorem to Eqn. 7.1 as follows to get the initial value of capacitor voltage:

$$\lim_{t \to 0} f(t) = \lim_{s \to \infty} sF(s) = \lim_{s \to \infty} s \times \frac{55.556 \times 10^4}{(s+1000)^2} \times \frac{1}{s} = \lim_{s \to \infty} \frac{55.556 \times 10^4}{(s+1000)^2} = 0$$

7.2.1.2 *Final Value Theorem*

Final value theorem could be applied to Eqn. 7.1 as follows to get the final value of capacitor voltage:

$$\lim_{t \to \infty} f(t) = \lim_{s \to 0} sF(s) = \lim_{s \to 0} s \times \frac{55.556 \times 10^4}{(s+1000)^2} \times \frac{1}{s}$$

$$= \lim_{s \to 0} \frac{55.556 \times 10^4}{(s+1000)^2} = 0.5556$$

One may use the following MATLAB commands to verify the values of step response obtained in Figure 7.2.

```
>> format short G
>> syms s t
```

```
>> ilaplace((55.5556e4/(s*(s+1000)^2)),s,t)
ans =
138889/250000 – 138889/250000*(1 + 1000*t)*exp(–1000*t)
>> t = 0;
>> 138889/250000 – 138889/250000*(1 + 1000*t)*exp(–1000*t)
ans =
0
>> t = 0.002;
>> 138889/250000 – 138889/250000*(1 + 1000*t)*exp(–1000*t)
ans =
0.3300
>> t = 0.1;
>> 138889/250000–138889/250000*(1 + 1000*t)*exp(–1000*t)
ans =
0.5556
```

A large signal average model of the buck converter using transfer function is shown in Figure 7.1. The step response illustrated in Figure 7.2 validates the accuracy of the model.

7.2.2 Performance analysis of the steady state model using state-space approach

One may use the state-space approach to obtain the steady state value of the capacitor voltage; the following algorithm could be used to obtain the steady state model of the buck converter; analysis of the results shows that the buck converter steps down an input voltage of 9 V to 5 V as illustrated in the output of the MATLAB program.

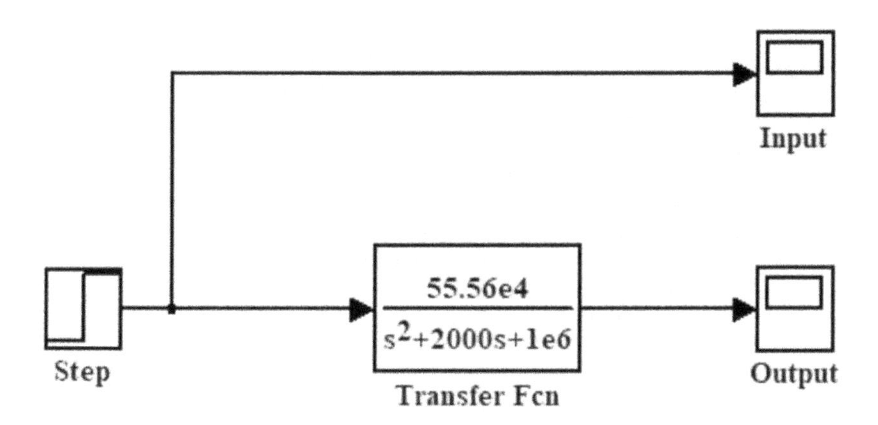

Figure 7.1 Large signal average model of a buck converter circuit using transfer function.

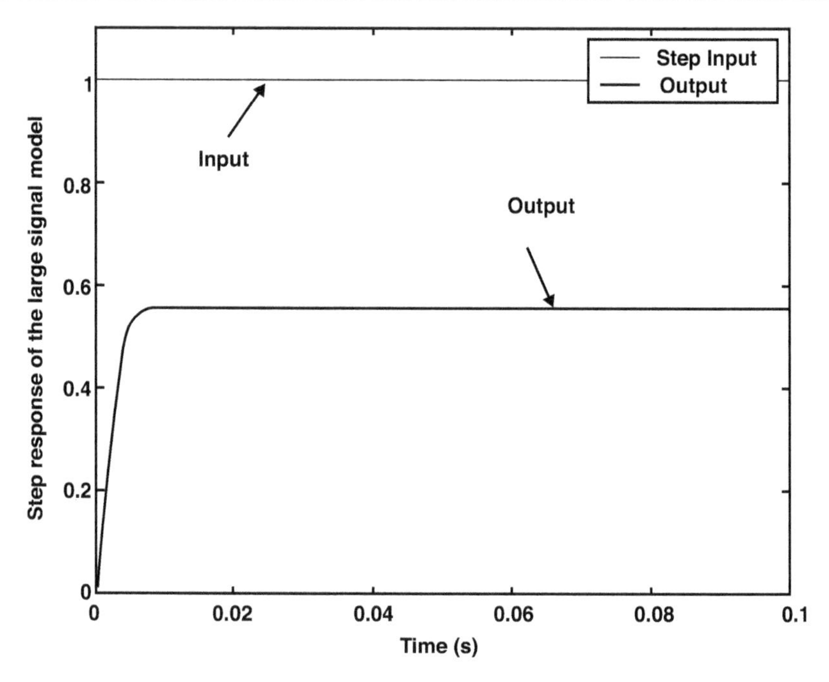

Figure 7.2 Step response of the large signal average model.

Algorithm

- Enter the values of input voltage (Vsbu), Duty cycle (Dsbu), Inductance (Lbu), Capacitance (Cbu), Load resistance (Rbu)
- Name the input vector, state vector and the output vector
- Calculate the parameter matrix of the steady state average model

$$[A_{sbust}] = \begin{bmatrix} 0 & \frac{-1}{L_{bu}} \\ \frac{1}{C_{bu}} & \frac{-1}{R_{bu}C_{bu}} \end{bmatrix}$$

- Calculate the input matrix of the steady state average model

$$[B_{sbust}] = \begin{bmatrix} \frac{D_{bu}}{L_{bu}} \\ 0 \end{bmatrix}$$

- Calculate the output matrix of the steady state average model

$$[C_{sbust}]' = [0 \;\; 1]$$

- Calculate the feed forward matrix of the steady state average model

$$[D_{sbust}] = [0]$$

- Calculate the output voltage

$$Vcbu = -C_{sbust} \times A_{sbust}^{-1} \times B_{sbust} \times V_{sbu}$$

- Print

$$[A_{sbust}], [B_{sbust}], [C_{sbust}], [D_{sbust}] \text{ and Vcbu}$$

With system specifications

Vsbu = 9 V; Dbu = 5/9; Lbu = 10 mH; Cbu = 100 μF; Rbu = 5 Ω; fbu = 1 kHz.

One may get the output of the MATLAB program as follows:

```
a =
          Iindbu         Vcbu
Iindbu  0              -0.00010
Vcbu    10000.00000    -2000.00000
b =
          Vsbu
Iindbu  5.55556e-005
Vcbu    0
c =
          Iindbu         Vcbu
Vcbu    0              1
d =
          Vsbu
Vcbu    0
Vcbu =
5
```

7.2.3 Performance analysis of the small signal average model using transfer function

The output response of the small signal average model of the buck converter with small signal variations in the in the duty cycle could be derived using transfer function model as follows:

For a unit step input Eqn. (6.79) could be written as:

$$\hat{V}_{cbu}(s) = \frac{9\times10^6}{(s+1000)(s+1000)}\times\frac{1}{s} \tag{7.12}$$

Using partial fractions Eqn. 7.12 could be written as:

$$\frac{9\times10^6}{s(s+1000)(s+1000)} = \frac{A_1}{s}+\frac{A_2}{(s+1000)}+\frac{A_3}{(s+1000)^2} \tag{7.13}$$

From Eqn. 7.13, one may get,

$$9\times10^6 = A_1\left(s^2+2000s+10^6\right)+A_2S(s+1000)+A_3S \tag{7.14}$$

From Eqn. 7.14, one may get,

$$9\times10^6 = 10^6A_1+s\left(2000A_1+1000A_2+A_3\right)+s^2\left(A_1+A_2\right) \tag{7.15}$$

From Eqn. 7.15, one may get,

$$9\times10^6 = A_1\times10^6 \therefore A_1 = 9 \tag{7.16}$$

From Eqn. 7.15 and Eqn. 7.16, one may get,

$$A_1+A_2 = 0 \therefore A_2 = -9 \tag{7.17}$$

From Eqn. 7.15, one may get,

$$2000A_1+1000A_2+A_3=0 \tag{7.18}$$

Substituting the values of A_1 and A_2 in Eqn. 7.18, one may get,

$$2000\times9+1000\times-9+A_3=0 \tag{7.19}$$

$$9000+A_3=0 \therefore A_3=-9000 \tag{7.20}$$

Substituting the values of A_1, A_2 and A_3 in Eqn. 7.13, one may get,

$$\hat{V}_{cbu}(s) = \frac{9\times10^6}{(s+1000)(s+1000)}\times\frac{1}{s} = \frac{9}{s}-\frac{9}{(s+1000)}-\frac{9000}{(s+1000)^2} \tag{7.21}$$

Taking inverse Laplace form of Eqn. 7.21, one may get,

$$\hat{v}_{cbu}(t) = L^{-1}\left(\frac{9}{s} - \frac{9}{(s+1000)} - \frac{9000}{(s+1000)^2}\right) \tag{7.22}$$

Applying inverse Laplace form to Eqn. 7.22, one may get,

$$\therefore \hat{v}_{cbu}(t) = 9 - 9e^{-1000t} - 9000te^{-1000t} \tag{7.23}$$

Substituting t = 0 s in Eqn. 7.23, one may get,

$$\hat{v}_{cbu}(t) = 9 - 9 = 0$$

Substituting t = 6 ms in Eqn. 7.23, one may get,

$$\hat{v}_{cbu}(t) = 8.8438V$$

Substituting t = 0.01 s in Eqn. 7.23, one may get,

$$\hat{v}_{cbu}(t) = 9V$$

One may also apply initial and final value theorems to calculate the initial value and final value of the capacitor voltage.

7.2.3.1 Initial value theorem

One may also apply initial value theorem to Eqn. 7.12 as follows to get the initial value of capacitor voltage:

$$\lim_{t\to 0} f(t) = \lim_{s\to\infty} sF(s) = \lim_{s\to\infty} s \times \frac{9\times10^6}{(s+1000)^2} \times \frac{1}{s} = \lim_{s\to\infty} \frac{9\times10^6}{(s+1000)^2} = 0$$

7.2.3.2 Final value theorem

Final value theorem could be applied to Eqn. 7.12 as follows to get the final value of capacitor voltage:

$$\lim_{t\to\infty} f(t) = \lim_{s\to 0} sF(s) = \lim_{s\to 0} s \times \frac{9\times10^6}{(s+1000)^2} \times \frac{1}{s} = \lim_{s\to 0} \frac{9\times10^6}{(s+1000)^2} = 9$$

Small signal average model of a buck converter using transfer function is shown in Figure 7.3. The step response illustrated in Figure 7.4 validates the accuracy of the model; one may note that for a duty cycle of 5/9, a multiplication factor of 5/9 should be taken into consideration in the output response.

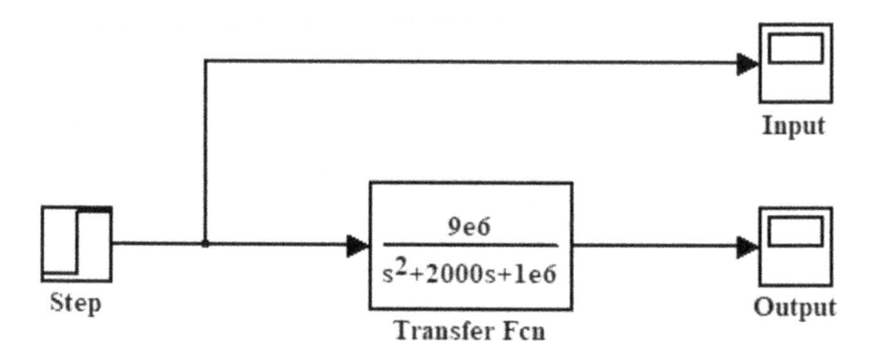

Figure 7.3 Small signal average model of a buck converter circuit using transfer function.

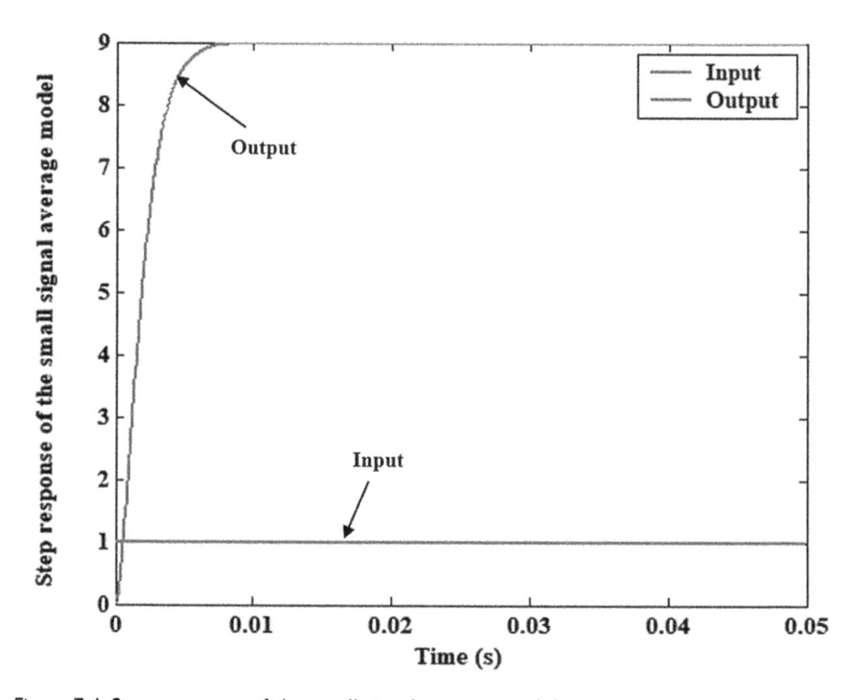

Figure 7.4 Step response of the small signal average model.

7.3 CLOSED LOOP CONTROL OF A BUCK CONVERTER USING ROOT LOCUS TECHNIQUE

One may use the root locus technique to decide the gain factor to achieve the desired output response independent of the parameter variations. The output response of a system can be slow and there can be steady state error and oscillations/overshoots in the output response. One may use a proportional controller to reduce the rise time; an integral controller to reduce the steady state error; a derivative controller to reduce oscillations/overshoots in the output response; or a combination of the controllers based on the specific requirements.

7.3.1 Closed loop control of the large signal average model of a buck converter

Consider a buck converter having the following specifications to regulate the output voltage to 3 V with a large variation in the input voltage of 20 V.

Vsbu = 20 V; Dbu = 0.8; Lbu = 10 mH; Cbu = 100 μF; Rbu = 5 Ω; fbu = 1 kHz.

An integral controller could be used to regulate the output of the buck converter with a large variation in the input voltage or the duty cycle. The stability of the system could be analyzed using the closed loop step response; for any instability, root locus technique could be used to vary the gain.

One may use the following algorithm to regulate the output of the large signal average model.

Algorithm

- Enter the values of input voltage (Vsbu), Duty cycle (Dsbu), Inductance (Lbu), Capacitance (Cbu), Load resistance (Rbu), Desired plant output (Pdo).
- Obtain the steady state average model of the buck converter
- Calculate the output voltage Vcbu
- Actual output Pao = Vcbu
- Error = Pdo – Pao
- Initialize the integral controller gain to unity (nc(s)/dc(s)) = 1/s
- Initialize feedback path transfer function to unity (nh(s)/dh(s)) = 1/1
- Obtain the transfer function model of the large signal average model
- Obtain the loop transfer function of the large signal average model
- While error not equal to zero
  - While key-in-input equal to continue
    - Plot the root locus of the loop transfer function
      - Get the values of gain ki from the root locus
      - Plot the closed loop step response
    - Enter the key-in-input
  - Display ki and plot the closed loop step response of the large signal model
  - Calculate the output of the large signal average model Vcbulr
  - Calculate Pao = Vcbu + Vcbulr
  - Calculate the new value of error = Pdo – Pao
- Display ki, error and Pao

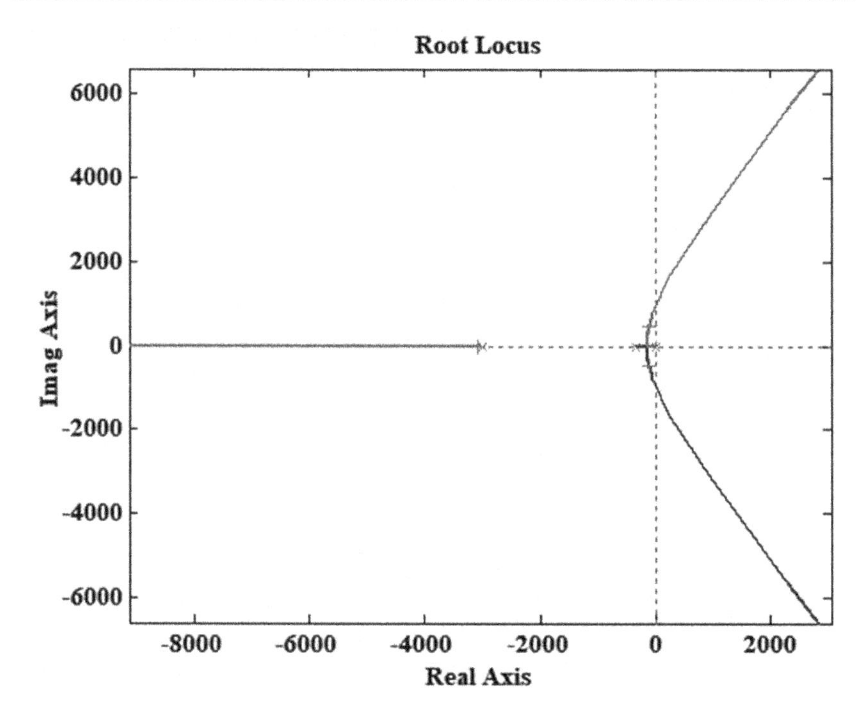

Figure 7.5 Root locus of the loop gain of the large signal average model with an integral controller.

One may note that the input of the large signal model is the error.

Different iterations have been conducted to decide the value of ki from the root locus of the large signal average model shown in Figure 7.5. The closed loop step response of the large signal average model of the system for a value of ki = 1633 is shown in Figure 7.6.

One may get the output of the steady state average model as follows:

a =

	Iindbu	Vcbu
Iindbu		−100.00000
Vcbu	10000.0000	−3333.33333

b =

	Vsbu
Iindbu	80.00000
Vcbu	0

c =

	Iindbu	Vcbu
Vcbu	0	1

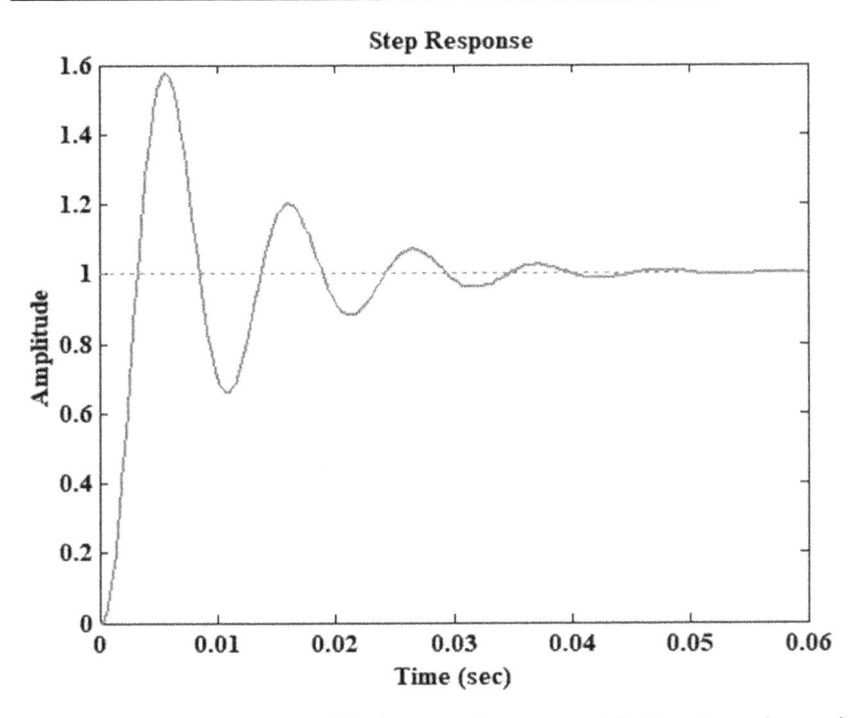

Figure 7.6 Closed loop response of the large signal average model with an integral controller of gain 1633.

```
d =
          Vsbu
Vcbu      0
Vcbu =
16
```

One may get the output of the large signal average model as follows:

```
Asbulr =
        1.0e + 004 *
                        0          -0.0100
                        1.0000     -0.3333
 Bsbulr =
              80          2000
              0           0
 Csbulr =
              0           1
 Dsbulr =
              0           0
```

For final selection of a point in the root locus for the determination of the gain of the integral controller, the values of the gain and location of the poles are as follows:

Select a point in the graphics window.

```
selected_point =
  -3.1462e + 003 - 2.0522e + 001i
  ki =
  1.6330e + 003
  p =
  1.0e + 003 *
  -3.1475
  -0.0929 + 0.6375i
  -0.0929 - 0.6375i
```

As the required output performance has been reached, one may quit the iterations; one may get the final values as follows:

```
0 to quit or 1 to continue0
  key-in-input =
  0
  Vcbulr =
  -13.0000
  Pao =
  3.000
  errorbu =
  0
```

7.3.2 Closed loop control of the small signal average model of a buck converter

Consider a buck converter having the following specifications to regulate the output voltage to 6V:

Vsbu = 9 V; Dbu = 0.556; Lbu = 10 mH; Cbu = 100 μF; Rbu = 5 Ω; fbu = 1 kHz.

One may use the small signal average model to regulate the output voltage with a proportional controller. The limiting value of the gain of the proportional controller kp could be determined as follows:

$$k_p G(j\omega) = \frac{9 \times 10^6 k_p}{s^2 + 2000s + 10^6} \tag{7.24}$$

One may note that at gain cross over frequency (ω_{gc})

$$k_pG(j\omega_{gc}) = \frac{9 \times 10^6 k_p}{(j\omega_{gc})^2 + 2000 \times j\omega_{gc} + 10^6} \tag{7.25}$$

For the system to be stable, at gain cross over frequency (ω_{gc}) the limiting value of Pm is zero.

$$0 = 180 + \angle G(j\omega_{gc})H(j\omega_{gc})k_p \tag{7.26}$$

$$180 = -\angle\left(\frac{9 \times 10^6 k_p}{-\omega_{gc}^{\ 2} + j2000\omega_{gc} + 10^6}\right) \tag{7.27}$$

$$180 = \tan^{-1}\left(\frac{2000\omega_{gc}}{10^6 - \omega_{gc}^{\ 2}}\right) \tag{7.28}$$

$$\left(\frac{2000\omega_{gc}}{10^6 - \omega_{gc}^{\ 2}}\right) = 0 \tag{7.29}$$

$$\omega_{gc} = 0\,{}^{rad}\!/_{sec} \tag{7.30}$$

At gain cross over frequency, gain in dB is zero, meaning that gain is 1; from Eqn. 7.25, the magnitude of the gain at gain cross over frequency,

$$\therefore \frac{9 \times 10^6 k_p}{\sqrt{\left((10^6 - \omega_{gc}^2)^2 + (2000\omega_{gc})^2\right)}} = \frac{9 \times 10^6 k_p}{10^6} = 1$$

$$\therefore k_p = 0.111$$

However, analysis of Figure 6.8 shows that gain margin is infinity, the gain can be increased to infinity. Figure 7.7 shows that both Gm and Pm are infinity for a gain of 0.05 less than the limiting value of kp; whereas Figure 7.8 shows that Pm is positive but not infinity for a gain of 0.15 greater than the limiting value of kp; one may note that the system is inherently stable.

One may use the following algorithm to determine the gain kp using root locus technique to regulate the output voltage of the buck converter with a small deviation in the duty cycle.

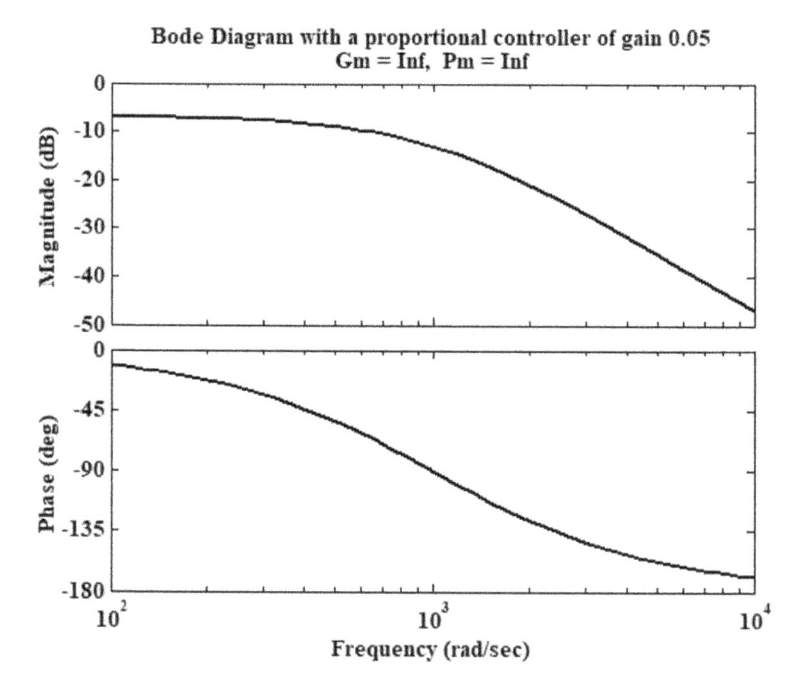

Figure 7.7 Bode diagram of the small signal average model with a proportional controller of gain 0.05.

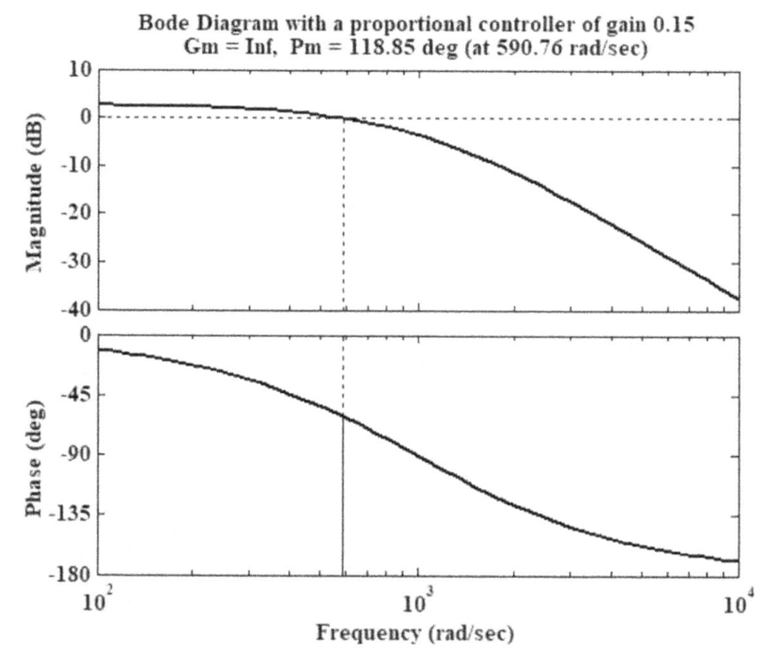

Figure 7.8 Bode diagram of the small signal average model with a proportional controller of gain 0.15.

Algorithm

- Enter the values of input voltage (Vsbu), Duty cycle (Dsbu), Inductance (Lbu), Capacitance (Cbu), Load resistance (Rbu), Desired plant output (Pdo).
- Obtain the state-space model of the steady state average model of the buck converter
- Calculate output voltage Vcbust
- Display Vcbust
- Actual output Pao = Vcbu
- Error = Pdo—Pao
- Initialize controller gain to unity (nc(s)/dc(s)) = 1/1
- Initialize feedback path transfer function to unity (nh(s)/dh(s)) = 1/1
- Obtain the transfer function model of the small signal average model
- Obtain the loop transfer function
- While error not equal to zero
  - Plot the root locus of the loop transfer function
  - While closed loop step response not stable
  - Get the values of gain from the root locus
  - Plot the output response of the steady state average model with new value of gain
- Display kp and Pao

Different iterations have been conducted to determine the value of kp from the root locus of the small signal average model shown in Figure 7.9.

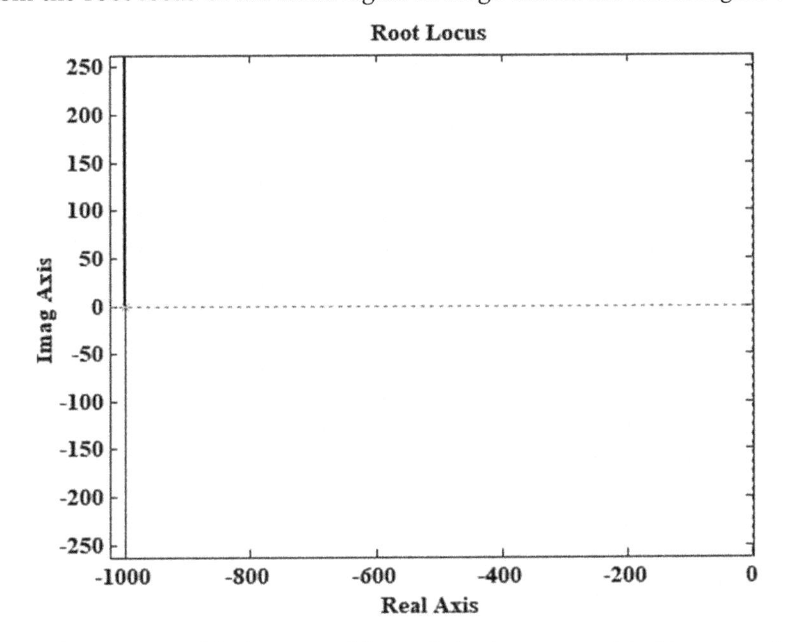

Figure 7.9 Root locus of the loop gain of the small signal average model with a proportional controller.

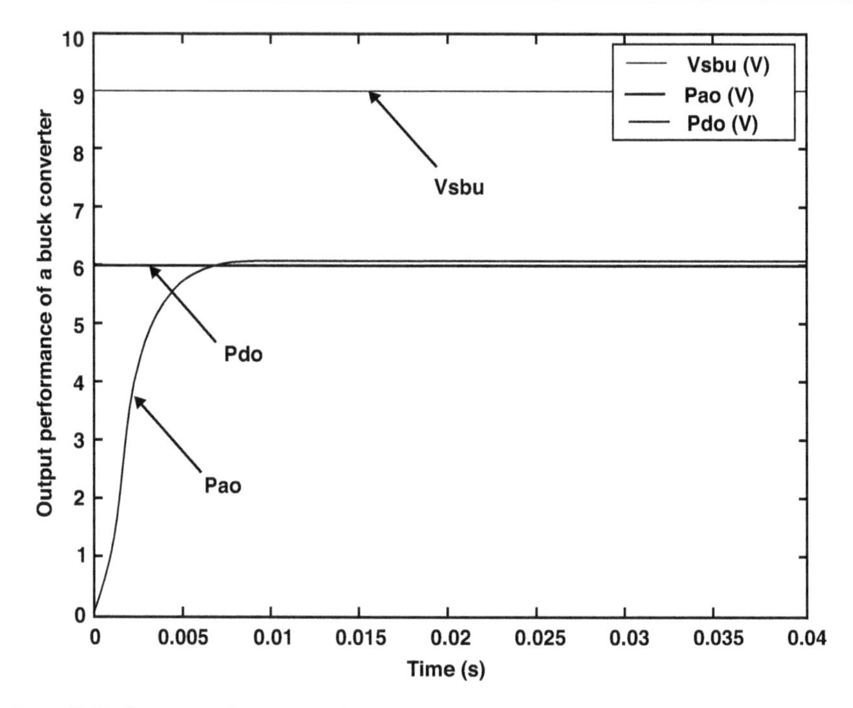

Figure 7.10 Output performance of a buck converter with a controller of gain 10.85.

For a value of kp = 10.85, the output voltage has been reached a steady state value of 6 V as shown in Figure 7.10.

One may get the output of the steady state average model as follows:

Vcbust =
6.0278
One may get the final results as follows:
kp =
10.8500
Pao =
6

BIBLIOGRAPHY

Benjamin C. K., Automatic control systems (5th ed.). India: Prentice Hall of India Private Limited, 1989.

Dorf R. C. and Bishop R. H., Modern control systems (1st ed.). World student series. USA: Addison-Wesley, 1998.

Electrical systems/basic alternating (AC) theory/transfer function analysis, online http://control.com/textbook/ac-electricity/transfer-function-analysis/, 2022.
Hadi S., Power system analysis (2nd ed.). Singapore: McGraw-Hill Education (Asia), 2004.
Manke B. S., Linear control systems with MATLAB applications (8th ed.). India: Khanna Publishers, 2005.
Nagoor K. A., Control systems (1st ed.). India: R B A Publications, 1998.
Ogata K., Modern control engineering (3rd ed., International). USA: Prentice Hall International Inc., 1997.
Rashid M. H., Tutorial on design and analysis of power converters. Malaysia: Universiti Putra Malaysia, 2002.
Umanand L., Switched mode power conversion, online http://nptel.ac.in/, 2014.

Chapter 8

Implementation of buck regulator systems

8.1 INTRODUCTION

As a reliable dc power supply, buck converters should provide regulated output voltages irrespective of the variations in the input, load and parameters restricted within the design factors. Conventional control techniques employ proportional-integral (PI) controllers along with pulse width modulation (PWM) technique to determine the duty cycle of the fully controllable switch. One may note that the PI controller generates the control voltage, which is then compared with a carrier signal switching at high frequency. However, it is possible to use a flip-flop-based control circuit to control the operation of the buck converters for stepping down the input voltage and for voltage regulation. In a flip-flop-based control circuit, switching signals are originated after comparing the control signals with a realistic fixed reference value; one may note that PWM control techniques do not employ a fixed reference value.

8.2 IMPLEMENTATION OF A BUCK REGULATOR SYSTEM USING PI-PWM CONTROL TECHNIQUE

Figure 8.1 depicts a buck regulator system using PI-PWM technique in MATLAB®/Power System Blockset environment with system specifications as seen in Table 8.1. One may note that the PI controller available in MATLAB/Simulink/Power System Blockset shown in Figure 8.2 generates the control voltage given by Eqn. 8.1.

$$v_{con} = k_p \text{ error} + k_i \int \text{error}$$

where

$$\text{error} = V_{rbu} - V_{cbu} \tag{8.2}$$

DOI: 10.1201/9781003511236-8

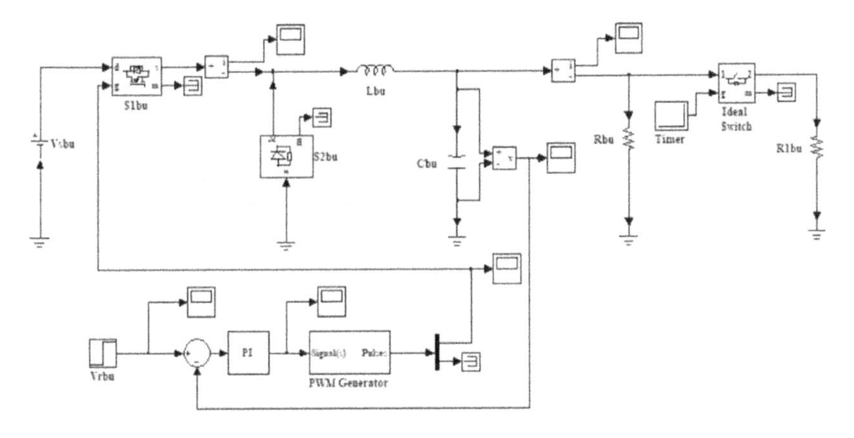

Figure 8.1 PSB model of a buck regulator system connected to two different loads.

Table 8.1 System Specifications of the Power Circuit of the PI-PWM Buck Regulator System

Input	*Inductance*	*Capacitance*	*Load resistance*	
V_{sbu} (V)	L_{bu} (mH)	C_{bu} (μF)	R_{bu} (Ω)	R_{1bu} (Ω)
9	10	100	5	5

One may use Zeigler-Nichols' trial and error method to determine the vales of the proportional constant k_p and the integral constant k_i. The control voltage v_{con} is then compared in a PWM generator available in the MATLAB/Simulink/Power System Blockset. Configuration of the PWM generator for a carrier frequency of 1 kHz is shown in Figure 8.3. The PWM module generates the switching pulses (sp) required to drive the fully controllable switch. One may introduce another resistance R_{1bu} in parallel with R_{bu} to vary the load; as shown in Figure 8.1, a timer controls the switching of R_{1bu}.

The performance of the overall system has been analyzed for different values of k_p and k_i. The system response shown in Figure 8.4 is for k_p=10 and k_i=20; one may note that even with a change in load (I_{Lbu}) at 0.05 s the controller is able to regulate the output voltage to the desired value of 5V. Logic levels of the switching pulses (sp) generated are illustrated in Figure 8.5; one may note that a positive pulse is generated when the actual output voltage (V_{cbu}) is less than the reference voltage (V_{rbu}) and the output of the PWM generator is held at zero when V_{cbu} is more than V_{rbu} to turn off the fully controllable switch.

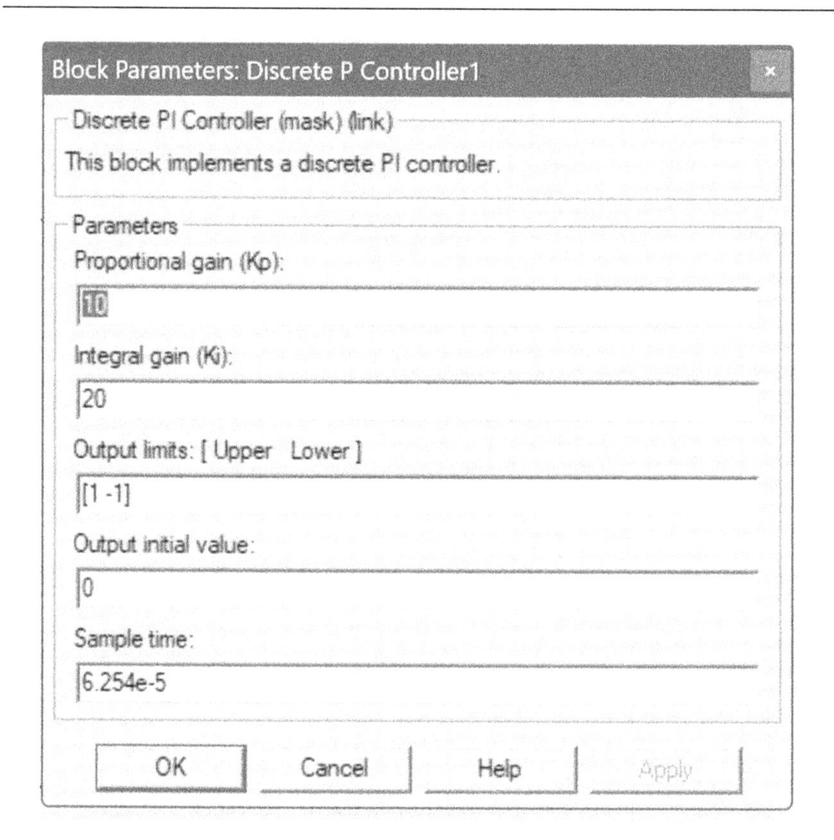

Figure 8.2 Settings of the PI controller.

8.3 A CASE STUDY ON A FLIP-FLOP CONTROLLED BUCK CONVERTER

As discussed in Chapter 5, a MOSFET plays a key role in the successful operation of a buck converter circuit. In a buck converter a MOSFET is used as a switch; an n-channel MOSFET can be in the ON state either by controlling the drain to source voltage (V_{ds}) with gate to source voltage (V_{gs}) more than the threshold value or by controlling V_{gs} with V_{ds} positive. One may note that depends on the availability of input and the output requirements one may opt for input voltage control or gate control.

8.3.1 High side vs low side switching configuration of an n-channel MOSFET

Figure 8.6 shows the high side switching configuration of an n-channel MOSFET having drain-source on state resistance R_{ds}. When the MOSFET is in the ON state, neglecting the voltage drop across R_{ds},

$$V_{se} \approx V_{de} = V_{ie} \tag{8.3}$$

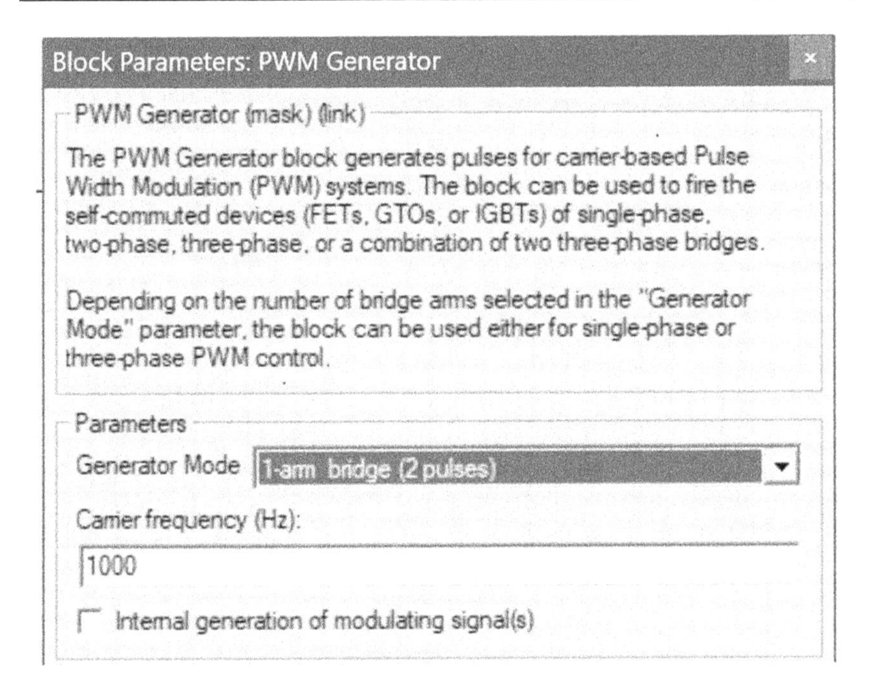
Block Parameters: PWM Generator

PWM Generator (mask) (link)

The PWM Generator block generates pulses for carrier-based Pulse Width Modulation (PWM) systems. The block can be used to fire the self-commuted devices (FETs, GTOs, or IGBTs) of single-phase, two-phase, three-phase, or a combination of two three-phase bridges.

Depending on the number of bridge arms selected in the "Generator Mode" parameter, the block can be used either for single-phase or three-phase PWM control.

Parameters

Generator Mode: 1-arm bridge (2 pulses)

Carrier frequency (Hz): 1000

Internal generation of modulating signal(s)

Figure 8.3 Settings of the PWM generator.

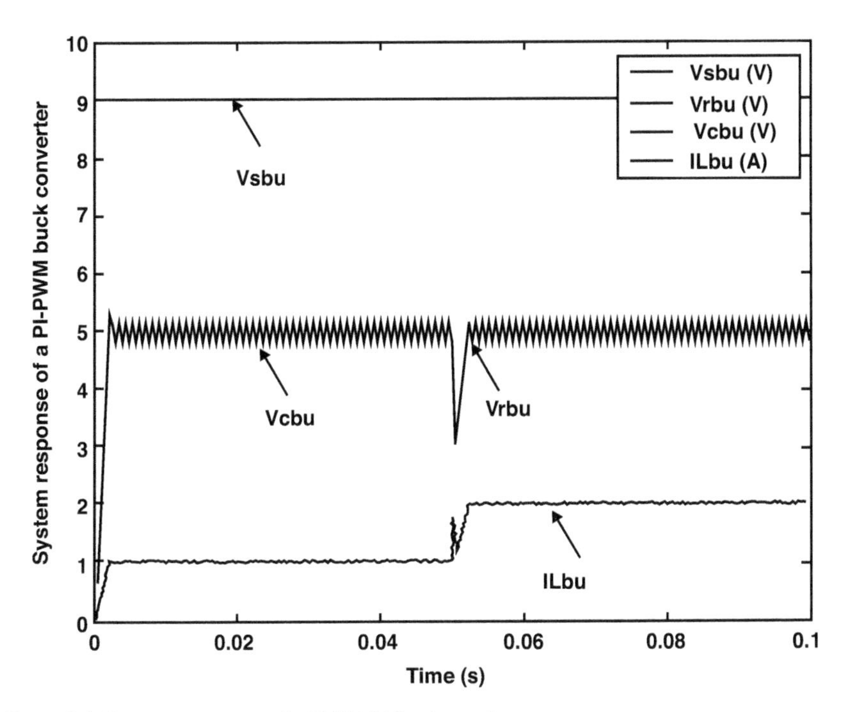

Figure 8.4 System response of a PI-PWM buck regulator system.

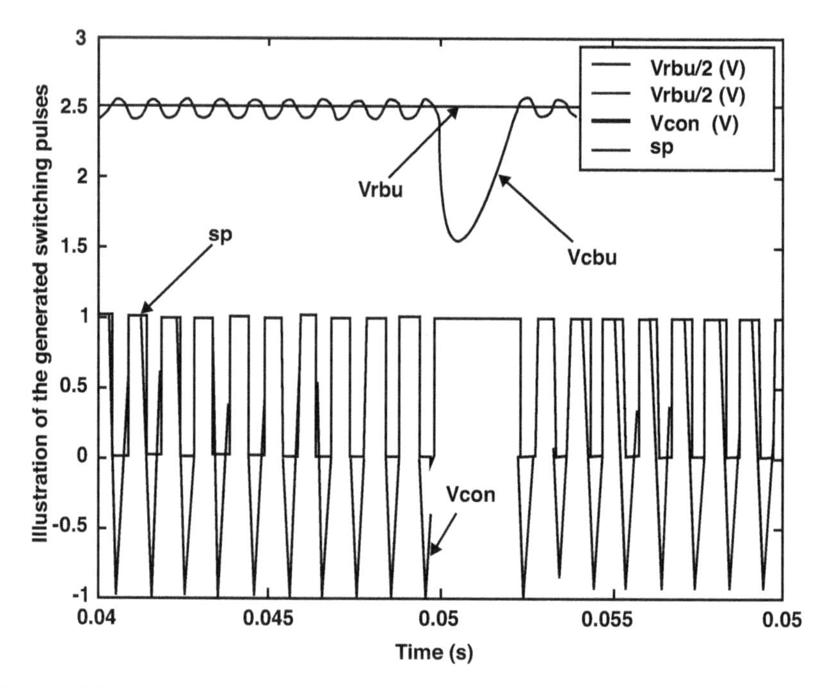

Figure 8.5 Switching pulses of a PI-PWM buck regulator system.

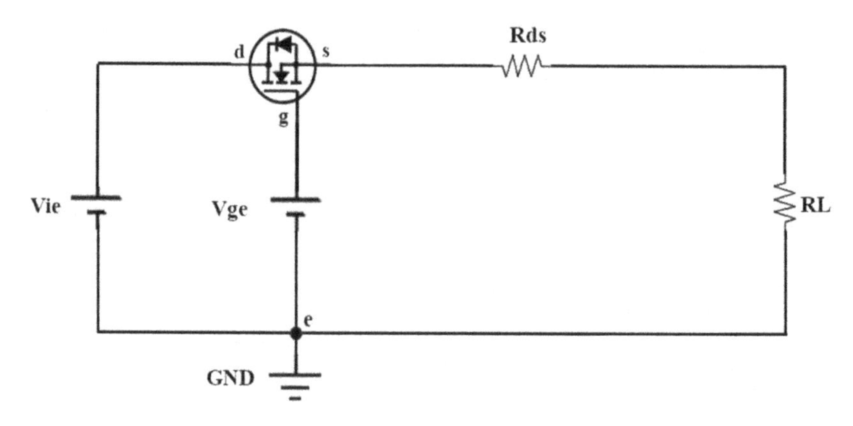

Figure 8.6 High side switching configuration of an n-channel MOSFET.

One may note that to turn on the MOSFET gate to source voltage V_{gs} should be higher than the threshold voltage V_{th}.
Meaning that

$$V_{ge} - V_{se} > V_{th} \tag{8.4}$$

$$V_{ge} > V_{ie} + V_{th} \tag{8.5}$$

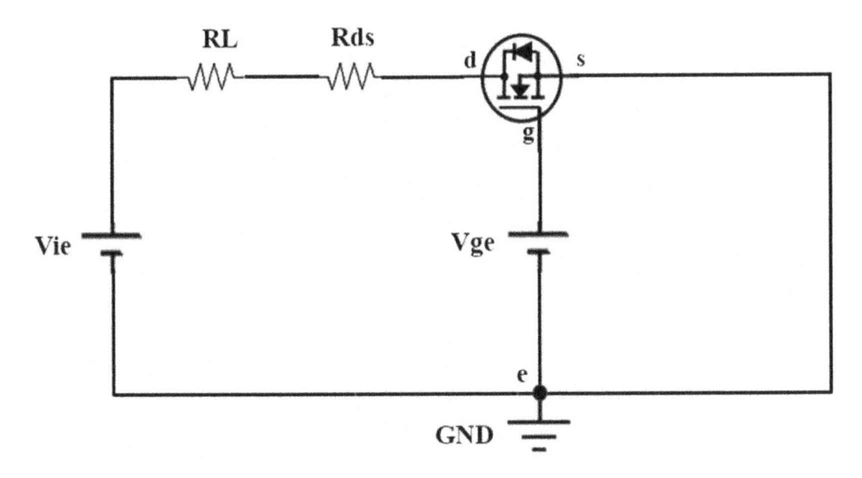

Figure 8.7 Low side switching configuration of an n-channel MOSFET.

Thus, for an input voltage V_{ie} of 5 V, with a V_{th} of 4 V, V_{ge} should be at least 9 V.

Figure 8.7 shows the low side switching configuration of an n-channel MOSFET. One may note that since source (s) side of the MOSFET is grounded,

$$V_{gs} = V_{ge} \tag{8.6}$$

Thus to turn on the MOSFET with a low side switching configuration,

$$V_{ge} > V_{th}$$

Thus independent of V_{ie}, for a V_{th} of 4 V, V_{ge} should be only 4 V.

8.3.2 Generation of 0–5 V regulated voltages from 5 V input power supply

Voltages in the range of 0–5 V find application in displays, illumination devices, sensor circuits, microprocessors and arduinos with digital electronics circuits. Linear voltage regulators and voltage dividers are commonly used to generate voltages in the range of 0–5 V; however the associated power dissipation and poor voltage regulation are the main constraints pertinent with these approaches.

The buck converter control circuit shown in Figure 8.8 adapts the ability of flip-flops to reproduce the input even at high frequency switching to generate voltages in the range of 0–5 V more efficiently. The digital signal generator of the control circuit in Figure 8.8 decides the input of the

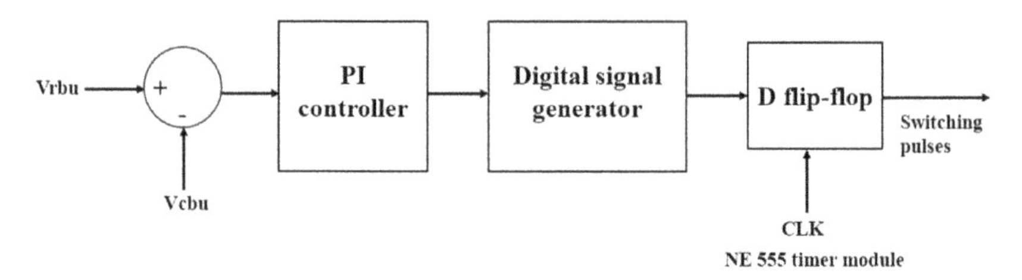

Figure 8.8 Control circuit.

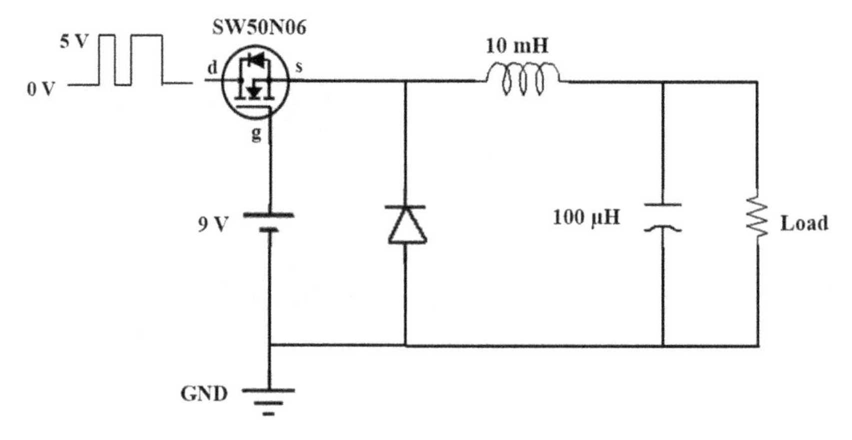

Figure 8.9 High side switching configuration of a buck converter.

SW50N06 MOSFET of the buck converter with high side switching configuration shown in Figure 8.9; one may note that voltage at the gate input is not changing and is connected to a fixed 9 V.

8.3.2.1 Power circuit of the buck converter

One may follow the power circuit configuration as shown in Figure 8.9 with SW50N06 n-channel MOSFET as the fully controllable switch.

Details of the terminals of SW50N06 are shown in Figure 8.10.

8.3.2.2 Control circuit

As shown in Figure 8.8, the output voltage of the buck converter V_{cbu} has been compared with the reference voltage V_{rbu} to determine the error voltage (error). The PI controller connected to the output of the comparator generates the required control voltage(v_{con}). If v_{con} is negative the digital signal generator outputs a logic zero and if v_{con} is positive the digital signal generator outputs a logic 1. The D flip-flop reproduces the output of the digital

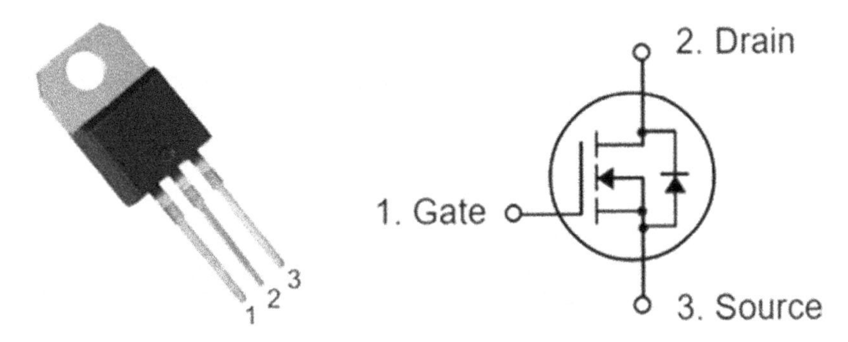

Figure 8.10 Pin diagram of SW50N06.

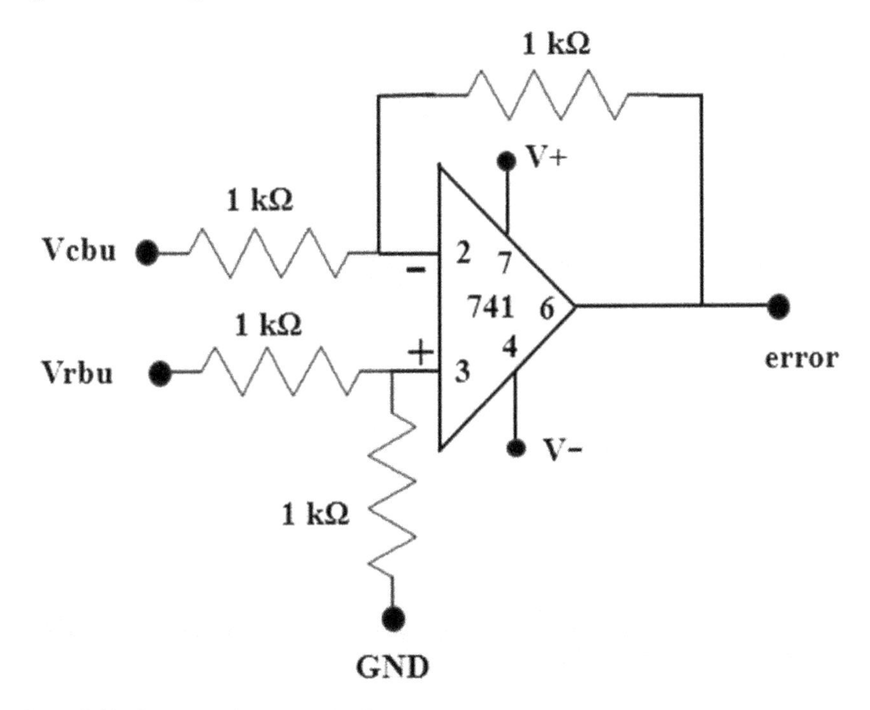

Figure 8.11 Op amp subtractor circuit.

signal generator at high frequency to supply input for the MOSFET of the buck converter.

8.3.2.2.1 Op amp subtractor circuit

An op amp circuit configured as a subtractor to generate the error voltage, the difference between the reference voltage V_{rbu} and the actual capacitor voltage V_{cbu}, is shown in Figure 8.11. Details of the terminals of 741 op amp

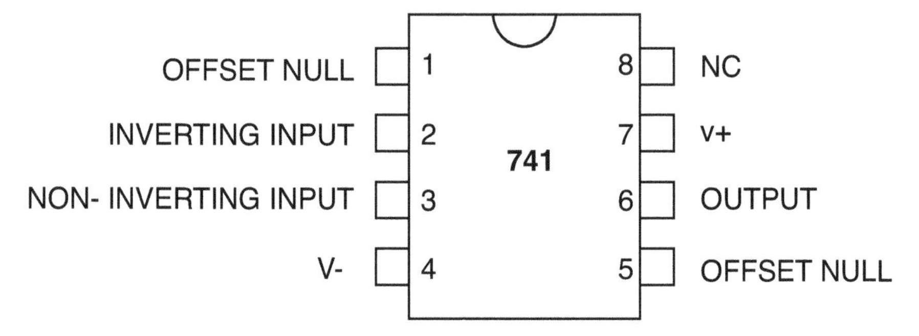

Figure 8.12 Pin diagram of 741.

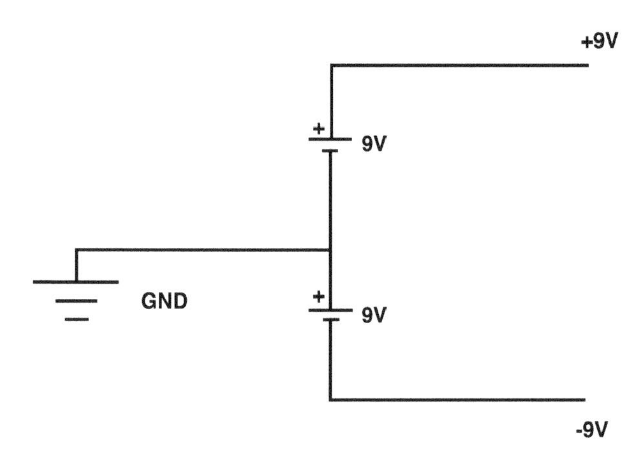

Figure 8.13 Power supply for the op amp.

are shown in Figure 8.12. If a dc power supply is unavailable, one may configure the batteries as shown in Figure 8.13 to generate the positive power supply (V+) and negative power supply (V-) for the op amp. One may note that with resistances of equal values, the output of the op amp circuit in Figure 8.11 is given by Eqn. 8.2.

8.3.2.2.2 PI controller

An op amp is configured as a PI controller as shown in Figure 8.14; one may note that the gain of the proportional controller k_p is given by Eqn. 8.3; and the gain of the integral controller k_i is given by Eqn. 8.4.

$$k_p = \frac{R_f}{R} \tag{8.8}$$

$$k_i = \frac{1}{RC} \tag{8.9}$$

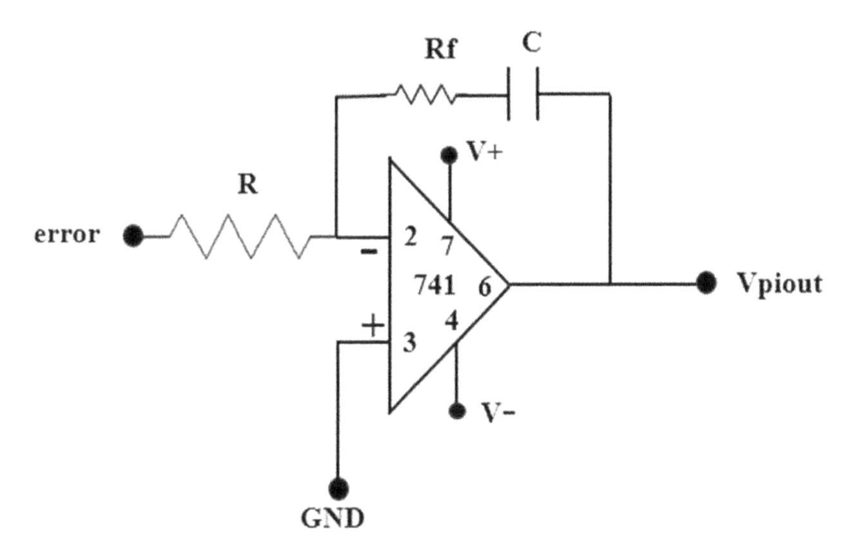

Figure 8.14 PI controller.

Taking $k_p = 10$ and $k_i = 20$, with $R = 1\ k\Omega$, one may get,
$R_f = 10\ k\Omega$ and $C = 50\mu F$.
One may note that the output of the PI controller shown in Figure 8.14 is given by Eqn. 8.10.

$$v_{piout} = -\left\{\frac{R_f}{R}\text{ error} + \frac{1}{RC}\int \text{error}\right\} = -v_{con} \tag{8.10}$$

8.3.2.2.3 D flip-flop

One may configure a 7474 D flip-flop as shown in Figure 8.15 to reproduce the output of the digital signal generator at high frequency to supply the input power of the MOSFET. The details of the terminals of 7474 dual positive edge triggered D flip-flops are shown in Figure 8.16. As shown in Figure 8.17, the timing diagram of a D flip-flop, one may note that the output of a positive edge triggered flip-flop waits for the positive edge of the clock (CLK) for any changes in the output response; and the output Q of a D flip-flop follows the D input at the positive edge of the clock. One may note that a logic 0 is equivalent to 0 volt and a logic 1 is equivalent to 5 V. If a dc power supply is unavailable, one may use the voltage regulator circuit shown in Figure 8.18 to provide dc supply for 7474. In the absence of a clock generator, one may use a 1 Hz–200 kHz NE 555 timer module shown in Figure 8.19; one may adjust the potentiometers available on the module to vary the frequency and duty cycle of the output clock signal.

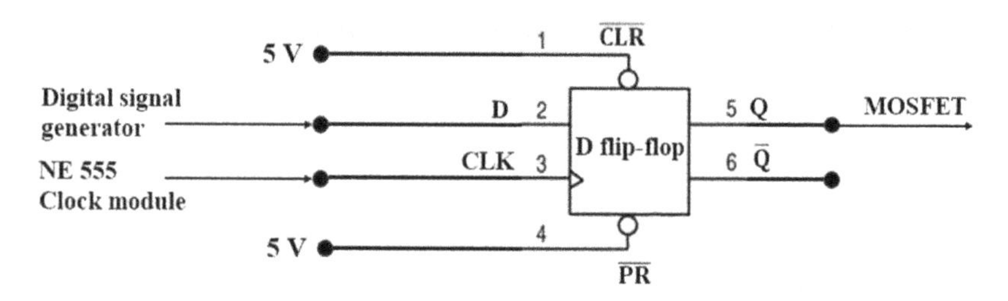

Figure 8.15 Configuration of the D flip-flop.

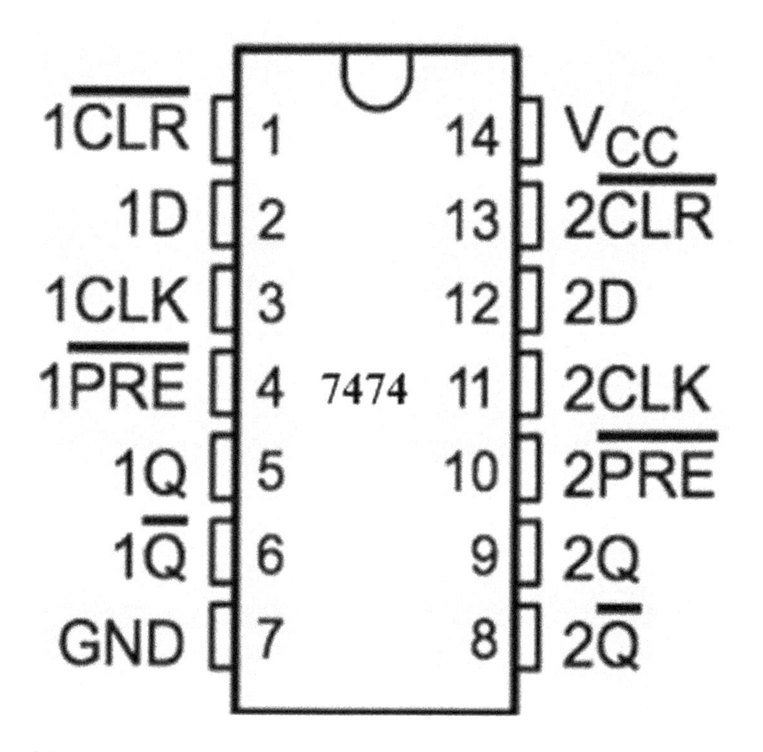

Figure 8.16 Pin diagram of 7474.

8.3.2.3 Results and analysis

The effectiveness of the system may be tested for voltages less than 5 V; batteries may be used as the reference sources. One may obtain the sample waveforms of the clock, D input and Q output as shown in Figure 8.20–Figure 8.22. One may note that the system is able to keep track of the reference voltages as shown in Figure 8.23 and Figure 8.24; with parameter variations one may adjust the values of controller gains to achieve the required output performance.

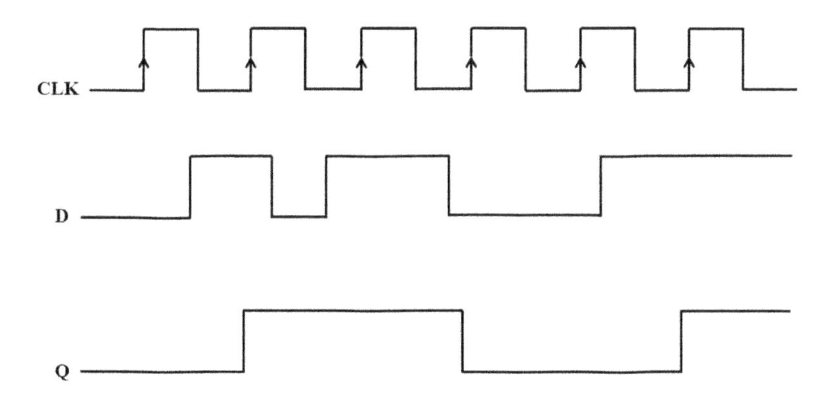

Figure 8.17 Timing diagram of a D flip-flop.

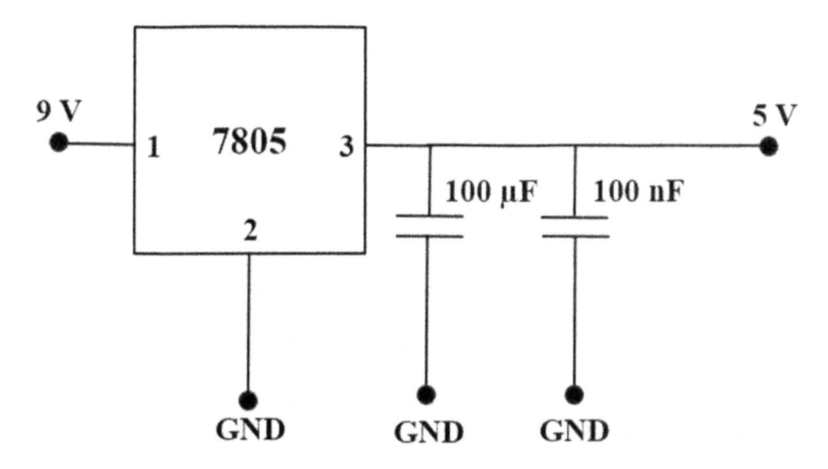

Figure 8.18 7805 9V to 5 V circuit.

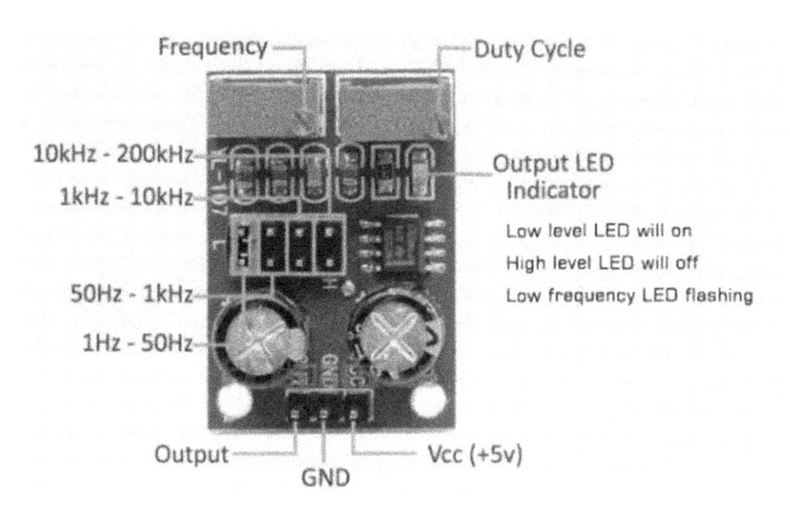

Figure 8.19 NE 555 timer module.

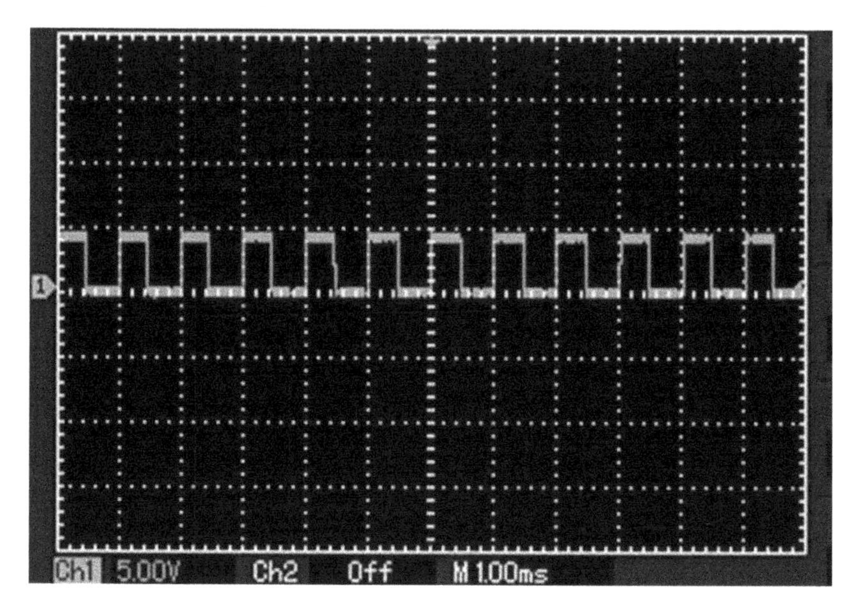

Figure 8.20 NE 555 1 kHz Clock signal.[1]

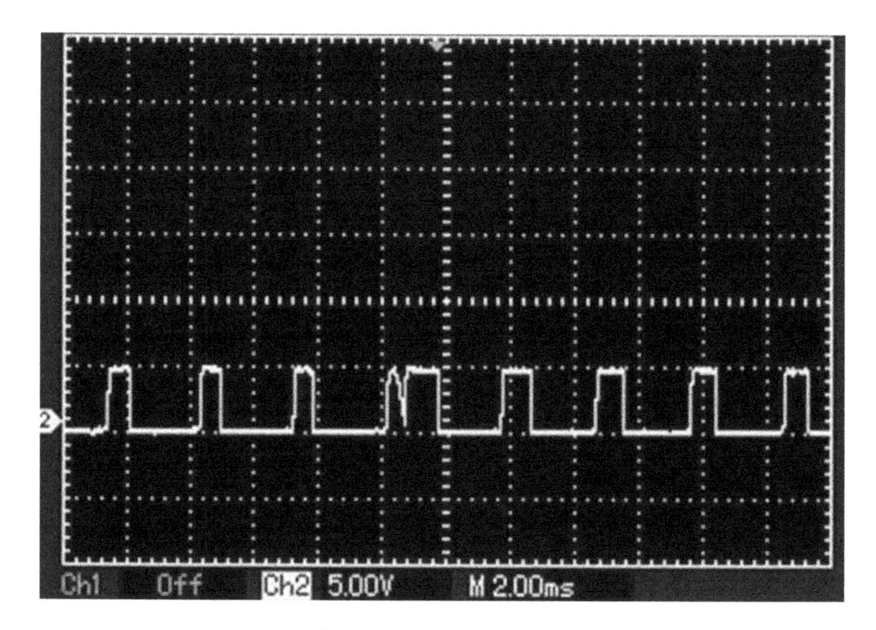

Figure 8.21 Waveform of input D.

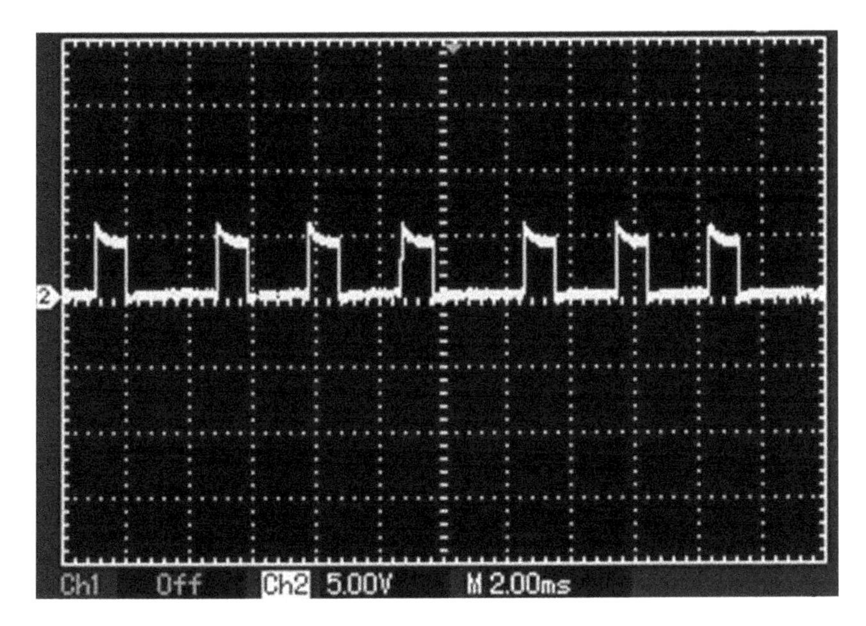

Figure 8.22 Waveform of output Q.

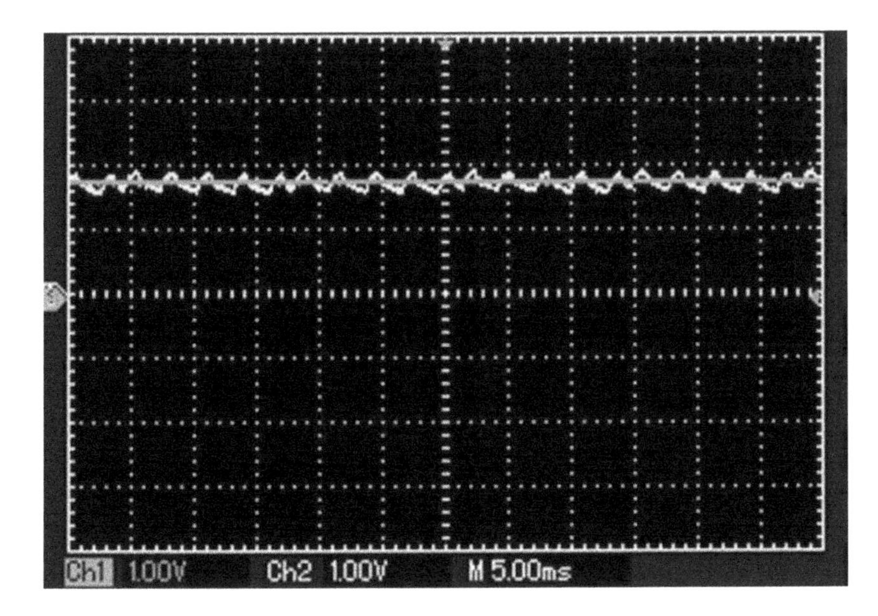

Figure 8.23 Waveform of V_{cbu} (ch2) for a V_{rbu} (ch1) of 1.6 V.

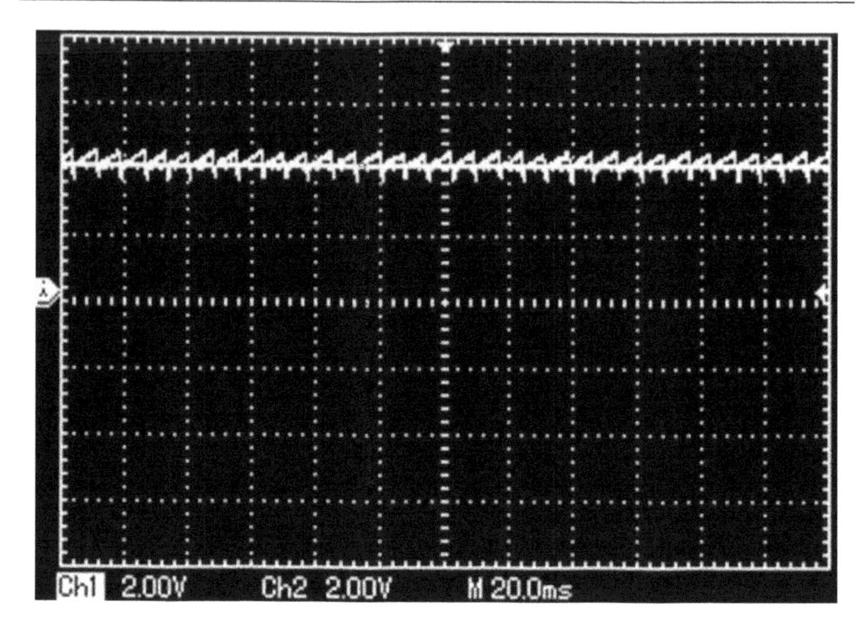

Figure 8.24 Waveform V_{cbu} (ch2) for a V_{rbu} (ch1) of 4 V.

8.3.3 A flip-flop controlled buck regulator system for solar panel applications

The scientific, technological and research advancement pave a way to overcome the demerits of generation of electricity from renewable energy resources. As a result, dependency on fossil fuel and other non-renewable energy resources has been reduced to establish a clean environment with less carbon and greenhouse gas emission. One may use rechargeable lithium-ion batteries to store energy from a solar panel; output of the solar panel may then be required to step down and regulate the battery voltages to make it useful for various applications.

8.3.3.1 Solar panel and battery charging controller module

One may use a Dragon 6 V, 100 mA, 0.6 epoxy solar panel[2] shown in Figure 8.25 to charge rechargeable lithium-ion battery to provide the input power for the buck converter.

CN3791 MPPT solar panel ion battery charging controller module[3] shown in Figure 8.26 shall be used to charge and protect the batteries. One may determine the number of the panels and batteries based on the power requirement.

Figure 8.25 Solar panel module.

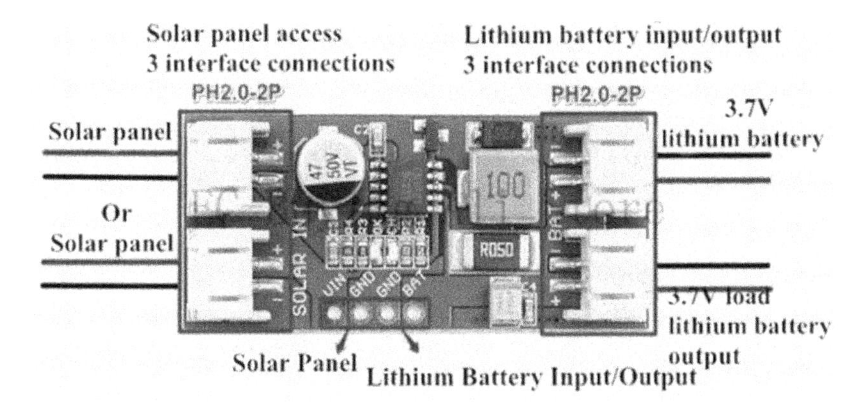

Figure 8.26 Battery charging controller module.

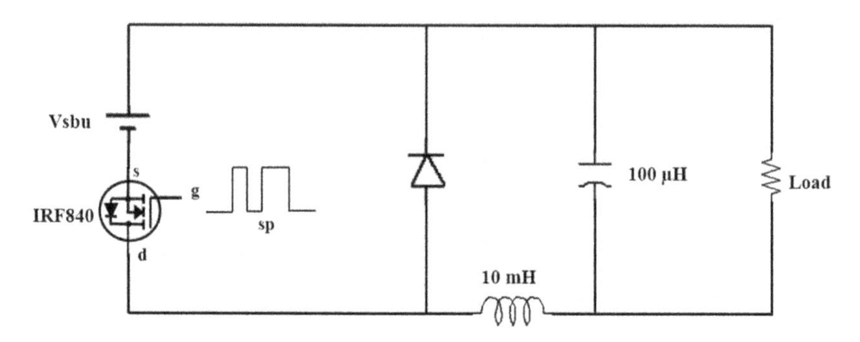

Figure 8.27 Low side switching configuration of a buck converter.

8.3.3.2 Power circuit of the buck converter

It should be noted that with a high side switching power circuit configuration in Figure 8.9, ground (GND) is the common lower potential terminal in the circuit wherever applicable and control circuit configuration is straight forward; one may note that however with a configuration as in Figure 8.9, the maximum output voltage is limited to 5 V. It should also be noted that the application of the buck converter explained in Section 8.3.2 is restricted within the power handling capability of the flip-flip and op amps.

To cover a wide range of applications, one may use low side switching configuration of the power circuit with source (s) terminal of the n-channel MOSFET grounded. One may note that by adapting the power circuit configuration in Figure 8.27, with proper control of the duty cycle, the system would be able to produce any voltages lower than the input voltages within the power handling capability of the power circuit elements.

One may note that the input voltage V_{sbu} should be higher than the required output voltage V_{cbu}; and irrespective of the irregularities, V_{sbu} can be from any source. One may use SW50N06/IRF840 as the fully controllable switch.

8.3.3.3 Control operation

One may use the control circuit in Figure 8.8 to generate the switching pulses of the MOSFET. One may note that with the power circuit configuration in Figure 8.27, on/off operation of the MOSFET is controlled by the duty cycle. It should be noted that unlike the power circuit configuration in Figure 8.9, as negative terminal of the capacitor is not grounded, careful attention should be paid to determine the actual capacitor voltage V_{cbu} using an op amp subtractor circuit.

8.3.3.4 Results and analysis

Two Dragon 6 V, 100 mA, 0.6 epoxy solar panels connected in series shall be used to charge the series combination of three 3.7 V rechargeable lithium-ion battery via the charging controller module CN3791. Fully charged batteries may have a total voltage of around 12 V and shall be used to provide the input supply V_{sbu} of the buck converter.

For V_{rbu} = 1.8 V; R_f = 1 kΩ; R = 100 Ω; C = 88 μF; R_{bu} = 1 Ω one may get the output voltage V_{cbu} (ch1) and output current I_L (ch2) as shown in Figure 8.28. One may use a battery to supply the reference voltage V_{rbu}. For V_{rbu} (ch1) = 5 V, one may get the output voltage V_{cbu} (ch2) as shown in Figure 8.29; one may use the 9 V to 5 V voltage regulator circuit in Figure 8.18 to generate the 5 V reference. It should be noted that to achieve the required output performance with variations in the reference voltage/load, the controller gains should be varied.

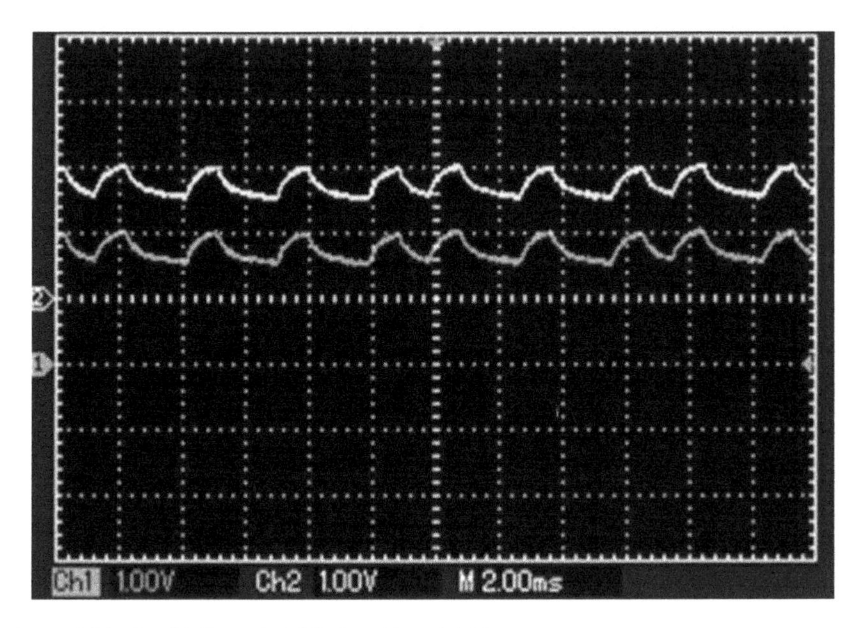

Figure 8.28 V_{cbu} (ch1) and I_L (ch2) for a V_{rbu} of 1.8 V.

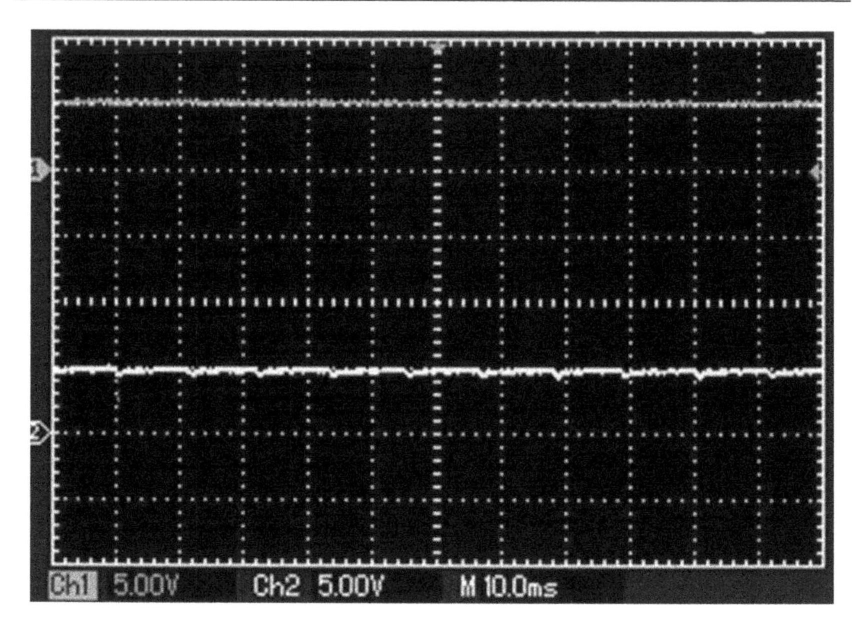

Figure 8.29 V_{cbu} (ch2) for a V_{rbu} (ch1) of 5 V.

NOTES

1 NE555 Timer module, ProtoSupplies (2022).
2 Epoxy solar panel, Lazada Malaysia (2022).
3 Solar panel ion battery charging controller module, Lazada Malaysia (2022).

BIBLIOGRAPHY

Barlet T. L. M., Industrial electronics: devices, systems and applications (1st ed.). USA: Delmar Publishers, 1997.

Benjamin C. K., Automatic control systems (5th ed.). India: Prentice Hall of India Private Limited, 1989.

Bimbra P. S., Power electronics (4th ed.). India: Khanna Publishers, 2006.

Bose B. K., Modern power electronics and AC drives (1st ed.). India: Pearson Education Inc., 2002.

Bradley D. A., Power electronics (2nd ed.). England: Chapman & Hall, 1995.

Dorf R. C. and Bishop R. H., Modern control systems (1st ed.). World student series. USA: Addison-Wesley, 1998.

Edminister J. A., Schaum's outline of theory and problems of electric circuits (Asian student ed.). Singapore: McGraw-Hill International Book Company, 1983.

Floyd T. L., Digital fundamentals (8th ed.). USA: Pearson Education Inc., 2003.

George M., Basu K. P., and Younis M. A. A., Digital simulation of static power converters using power system blockset (1st ed.). Germany: LAP LAMBERT Academic Publishing, 2012.

Hambley R. A., Electrical engineering: principles & applications (3rd ed., International). USA: Pearson Education International, 2005.

Hughes E., Electrical and electronics technology (10th ed.). UK: Pearson Education Ltd., 2008.

Kissell T. E., Industrial electronics: applications for programmable controllers, instrumentation and process control, and electrical machines and motor controls (2nd ed.). USA: Prentice-Hall Inc., 2000.

Lazada Malaysia, Epoxy solar panel, online https://lazada.com.my/, 2022a.

Lazada Malaysia, Solar panel ion battery charging controller module, online https://lazada.com.my/, 2022b.

Maloney T. J., Modern industrial electronics (4th ed.). USA: Prentice-Hall Inc., 2001.

Malvino A. P. and Leach D. P., Digital principles and applications (4th ed.). Singapore: Mc-Graw Hill Co., 1969.

Manke B. S., Linear control systems with MATLAB applications (8th ed.). India: Khanna Publishers, 2005.

Mehta V. K. and Mehta R., Principles of electrical engineering and electronics (multicolour illustrative edition). India: S. Chand & Company Ltd., 2007.

Mehta V. K. and Mehta R., Principles of electronics (multicolour illustrative edition). India: S. Chand & Company Ltd., 2006.

Millman J. and Halkias C., Integrated electronics (1st ed.). Singapore: Mc Graw-Hill International Editions, 1977.

Mitchell F. H. Jr. and Mitchell F. H. Sr., Introduction to electronics design (2nd ed.). USA: Prentice-Hall International Inc., 1992.

Mohan N., Power electronics modeling simplified using PSPICE (Release 9). Canada: University of Minnesota, 2002.

Mohan N., Undeland T. M., and Robbins W. P., Power electronics: converters, applications and design (3rd ed.). USA: John Wiley & Sons Inc., 2003.

Nagoor K. A., Control systems (1st ed.). India: R B A Publications, 1998.

Ogata K., Modern control engineering (3rd ed., International). USA: Prentice Hall International Inc., 1997.

Petruzella F. D., Industrial electronics (1st ed.). Electrical & electronic technology series. Singapore: McGraw-Hill International Editions, 1996.

ProtoSupplies, NE555 Timer module, online https://protosupplies.com/, 2022.

Rai, H. C., Industrial and power electronics (1st ed.), India: Umesh Publications, 1987.

Rashid M. H., 2 days course on power electronics and its applications. Malaysia: Universiti Putra Malaysia, 2004.

Rashid M. H., Power electronics circuits, devices and applications (3rd ed.). India: Pearson Education Inc., 2004.

Rashid M. H., Tutorial on design and analysis of power converters. Malaysia: Universiti Putra Malaysia, 2002.

Sen P. C., Principles of electric machines and power electronics (2nd ed.). USA: John Wiley & Sons Inc., 1999.

Singh A. and Triebel W. A., The 8086 and 80286 microprocessors: hardware, software and interfacing. USA: Prentice-Hall Inc., 1990.

Subramanyam V., Power electronics (1st ed.). India: New Age International Private Ltd., 2003.
Texas instruments, online www.ti.com/, 2022.
The MathWorks Inc., Power system blockset for use with Simulink. USA: MathWorks Incorporated, 2000.
Theraja A. K. and Theraja B. L., A textbook of electrical technology (25th ed.). India: S. Chand & Company Ltd., 2008.
Tocci R. J., Widmer N. S., and Moss G. L., Digital systems principles and applications (10th ed.). USA: Pearson Education Inc., 2007.
Umanand L., Switched mode power conversion, online http://nptel.ac.in/, 2014.
Wildi T., Electrical machines, drives, and power systems (4th ed.). USA: Prentice Hall International Inc., 2000.

Index

Q

R

S

For Product Safety Concerns and Information please contact our EU
representative GPSR@taylorandfrancis.com
Taylor & Francis Verlag GmbH, Kaufingerstraße 24, 80331 München, Germany

www.ingramcontent.com/pod-product-compliance
Ingram Content Group UK Ltd.
Pitfield, Milton Keynes, MK11 3LW, UK
UKHW021122180425
457613UK00005B/184

* 9 7 8 1 0 3 2 6 2 7 7 3 1 *